THE ECONOMICS OF BIOFUEL POLICIES

PALGRAVE STUDIES IN AGRICULTURAL ECONOMICS AND FOOD POLICY

Series Editor: Christopher B. Barrett, Cornell University, USA.

Agricultural and food policy lies at the heart of many pressing societal issues today, and economic analysis occupies a privileged place in contemporary policy debates. The global food price crises of 2007–2008 and 2010–2011 underscored the mounting challenge of meeting rapidly increasing food demand in the face of increasingly scarce land and water resources. The twin scourges of poverty and hunger quickly resurfaced as high-level policy concerns, partly because of food price riots and mounting insurgencies fomented by contestation over rural resources. Meanwhile, agriculture's heavy footprint on natural resources motivates heated environmental debates about climate change, water and land use, biodiversity conservation, and chemical pollution. Agricultural technological change, especially associated with the introduction of genetically modified organisms, also introduces unprecedented questions surrounding intellectual property rights and consumer preferences regarding credence (i.e., unobservable by consumers) characteristics. Similar new agricultural commodity consumer behavior issues have emerged around issues such as local foods, organic agriculture, and fair trade, even motivating broader social movements. Public health issues related to obesity, food safety, and zoonotic diseases such as avian or swine flu also have roots deep in agricultural and food policy. And, agriculture has become inextricably linked to energy policy through biofuels production. Meanwhile, the agricultural and food economy is changing rapidly throughout the world, marked by continued consolidation at both farm production and retail distribution levels, elongating value chains, expanding international trade, and growing reliance on immigrant labor and information and communications technologies.

In summary, a vast range of topics of widespread popular and scholarly interest revolve around agricultural and food policy and the economics of those issues. This series features leading global experts writing accessible summaries of the best current economics and related research on topics of widespread interest to both scholarly and lay audiences.

The Economics of Biofuel Policies: Impacts on Price Volatility in Grain and Oilseed Markets
by Harry de Gorter, Dusan Drabik, and David R. Just

THE ECONOMICS OF BIOFUEL POLICIES

IMPACTS ON PRICE VOLATILITY IN GRAIN AND OILSEED MARKETS

Harry de Gorter, Dusan Drabik, and David R. Just

palgrave
macmillan

THE ECONOMICS OF BIOFUEL POLICIES

First published in 2015 by
PALGRAVE MACMILLAN®
in the United States—a division of St. Martin's Press LLC,
175 Fifth Avenue, New York, NY 10010.

Where this book is distributed in the UK, Europe and the rest of the world, this is by Palgrave Macmillan, a division of Macmillan Publishers Limited, registered in England, company number 785998, of Houndmills, Basingstoke, Hampshire RG21 6XS.

Palgrave Macmillan is the global academic imprint of the above companies and has companies and representatives throughout the world.

Palgrave® and Macmillan® are registered trademarks in the United States, the United Kingdom, Europe and other countries.

ISBN: 978–1–137–41484–7

Library of Congress Cataloging-in-Publication Data

De Gorter, Harry, author.
The economics of biofuel policies : impacts on price volatility in grain and oilseed market / Harry de Gorter, Dusan Drabik, David R. Just.
pages cm.—(Palgrave studies in agricultural economics and food policy)
Includes bibliographical references and index.
ISBN 978–1–137–41484–7 (hardback : alk. paper)
1. Biomass energy—Prices—Government policy. 2. Biomass energy—Economic aspects. I. Drabik, Dusan, author. II. Just, David R., author. III. Title. IV. Series: Palgrave studies in agricultural economics and food policy.

HD9502.5.B542D44 2015
338.1′331—dc23 2014040145

A catalogue record of the book is available from the British Library.

Design by Newgen Knowledge Works (P) Ltd., Chennai, India.

First edition: April 2015

10 9 8 7 6 5 4 3 2 1

To my mom, the true intellectual in the family
To my parents
To my father who taught me to think and learn for myself

CONTENTS

FIGURES

TABLES

BOXES

FOREWORD

There are at least three reasons why it is a great pleasure to write the foreword for this volume. First, this is the inaugural volume in an important new book series that Palgrave Macmillan is launching on Agricultural Economics and Food Policy. The broad topics of this new series, which I am privileged to edit, lie at the heart of many of today's most pressing societal issues, as has been vividly underscored by the global food price crises of 2007–2008 and 2010–2011, the resulting food riots that engulfed dozens of nations, and the subsequent high-level calls for renewed investment in agriculture and especially food systems.

Second, arguably, no topic in agricultural economics and food policy has had a more profound impact on society, and in the research community, over the past decade than has biofuels policy. The introduction of unprecedentedly large mandates for biofuels—ethanol in the United States, in particular—precipitated an explosion of academic articles debating the impacts of those policies and considering how these policies might be reformed to improve social welfare. Biofuels policy has had profound impacts on agricultural producers and consumers worldwide, both directly through product innovation and pricing, and indirectly, through induced innovations in other energy sources, compensatory policy measures, and adjustments to the political economy of contemporary food, energy, and agricultural policies. Understanding these impacts is key to making sense of commodity price fluctuations, trade in fuel and food, energy markets, and efforts to mitigate climate change, among other topics. But, a tremendous amount of confusion persists as biofuels policies remain widely misunderstood.

Third, there are no better qualified scholars to help readers make sense of this highly controversial subject than Harry de Gorter, Dusan Drabik, and David Just. Their various papers on the economics of biofuels are, by far, the most cited in the mass of research done over the past decade or so on this important topic. They bring rigorous theorizing and careful empirical analysis to bear in a way that helpfully integrates diverse strands of the literature. I know this from firsthand observation, as Harry and David are colleagues of mine in the Charles H. Dyson School of Applied Economics and Management at Cornell University, and Dusan was a star graduate student here. Their thoughtful and meticulous work has corrected some of my own prior misconceptions about biofuel issues. By confronting some of the oversimplifications and mistaken inferences of the early literature on the topic, these authors have done much to bring the full analytical force of contemporary economics to bear on a topic of historic importance. They have forced us to move beyond toy models that represent mandates or subsidies as simple shifts in supply and demand—and ignore the actual production process which returns some byproduct to the corn market, the incentives induced by firms' profit objectives, by the constraints of competition in ethanol production, or by fuel consumers' motives in product choice. For those of us more inclined toward empirical analysis, they have helpfully called attention to misspecification in time series estimation of relationships between the various prices of corn, ethanol, gasoline, and oil that fail to take into proper account the state-conditional pricing regimes necessarily implied by policy structure. In short, Harry, Dusan, and David have educated me a great deal about the economics of biofuels. That is why I asked them to write this book.

The model they present in this volume is not only simple and comprehensible, but, at least as importantly, it is based upon the right fundamentals. They augment tried-and-true models of commodity demand, supply, and storage with the key physical and behavioral relationships that define the current biofuels policies regime. In their model, ethanol producers seek to maximize profits in a fully competitive industry, consumers minimize transport costs, and the physical outputs from the production process are accounted for fully. Into this rigorous, robust model, they introduce current policy designs and trace out their implications, especially for commodity prices. They highlight how the layering of one policy on top of another often offsets the impacts of each policy on its own, dampening and distorting effects. Determining when each policy is and is not effective is one key to unpacking the impacts of biofuels policies and to help design sensible and impactful policies.

The primary goal of this book is to provide a point of departure for those embarking on new research on the economics of biofuels. This remains a very young field, and surely much remains to be uncovered. However, the unsurpassed expertise that de Gorter, Drabik, and Just bring to this topic, and the rigor and clarity with which they disentangle the complex web of interconnections between different market actors and policy instruments, makes this book the right place to start in any serious new inquiry into the economics of biofuels. Just as I have learned from these expert authors, so will readers of this book be rewarded with clearer insights on the evolving biofuels economy.

This is surely one of the most important topics in agricultural economics and food policy in the early twenty-first century and a topic of monumental influence over many key societal concerns today, be it climate change, food prices, hunger, land use, or renewable energy. It is a great pleasure to invite you to read, learn from, and enjoy this important new book.

Christopher B. Barrett
Cornell University

PREFACE

In the second half of 2007 and the first half of 2008, the world witnessed a spectacular boom in cereal and oilseed prices although the US economy was in a recession. *The Economist* (2008a,b) noted that this boom caught everybody by surprise; it was a "silent tsunami." All grain and oilseed prices tripled. The rice price tripled in a year; doubling in the first four months of 2008 alone. Policymakers in developed and developing countries alike appeared ill-equipped to deal with high prices and the implications for the world's poor, and with the ensuing price volatility. All prices fell precipitously during the Great Recession, yet corn prices bottomed at $3.30 per bushel in September 2009, far above the $1.62 per bushel low in November of 2005 when the world enjoyed its biggest economic boom ever. Nevertheless, virtually all economists declared it was yet another "boom and bust" cycle, expecting a coming "mean reversion" in food commodity prices, just as before.

But cereal and oilseed prices overtook their 2008 peak in 2010–2011 and corn prices peaked even higher in 2012–2013. The price level and volatility were unprecedented and continue to be so at the time of this writing. The cereal and oilseed markets seem to be in constant turmoil with a series of "food crises" as the price for basic food commodities was atop the G-20 agenda in 2008 and again in 2011. Others deemed it was not really a "food crisis" as developing country policy response insulated most poor consumers from high prices in 2008. Nevertheless, the global food economy is in the middle of a major price reversal, after a several-decade long decline in real prices going into 2006.

Why do we say "going into 2006" when economists, including *The Economist*, declared the price boom a sudden and surprising "silent tsunami" that first emerged in the latter part of 2007 and peaked in July 2008? Because for six months beginning in September 2006, the Central Illinois farm price of corn rose 88 percent (Kansas City white corn price up 107 percent), culminating in the Mexican tortilla crisis in January 2007 and India's wheat export ban in February 2007—the first of many developing countries' policy responses to come. Why? Because as corn–ethanol production capacity was doubling in 2006, high ethanol and crude oil prices created a link between corn and ethanol prices for the first time ever.

Why were ethanol prices high? This was due to biofuel policies, and particularly US environmental regulations that controlled the use of fuel additives, making ethanol an attractive economic alternative. Furthermore, US energy policy in 1978 established a federal 51 cent per gallon blenders' tax credit for ethanol that had been dormant all this time due to the low value of ethanol as a fuel. However, high crude oil prices activated this tax credit by creating a link between crude oil (gasoline) and ethanol prices.

The most important aspect of all of this is that the "multiplier" between corn and ethanol prices is very high: for every one cent per gallon increase in the price of ethanol, the price of corn rose four cents per bushel (this multiplier reached a peak of 4.82 in December 2013). This six-month time period beginning in September 2006 that shook the world (as we observed it) actually did not shake the world in the mind of economists at the time! Economists focused instead on bad weather that reduced world wheat production a whopping 3.1 percent in 2006–2007 (total grains and oilseed production was essentially unchanged). What we regarded as a sharp increase in corn prices in 2006–2007 was not a shot heard around the world when it should have been. Why? Because nobody understood it for what it was! But how did we know? Well, that is the purpose of this book: to explain and extend the basic model of, for a lack of a better term, "the theory of biofuel policy" that we originally published in four working papers in 2007 (de Gorter and Just 2007a, b, c, d)[1] and emailed at the time to approximately 500–600 economists around the world (some deny they received it; others complained they had!)

Although these papers analyzed a far-ranging array of topics, including the welfare economics and environmental effects of several biofuel policies, international trade, and a host of other issues, front row and center in these papers were the corn–ethanol price link and the very high multiplier effect on corn prices. If one looks closely at the last column of Table 2 in de Gorter and Just (2007d, 2008b) for the year 2006–2007, the ethanol price premium (or corn price increase due to biofuel policies) is 97 percent of the corn price or $2.95 per bushel. The corn price was $3.05 per bushel.[2]

In early March 2008, Don Mitchell at the World Bank invited us to give a seminar on all of this. In the course of discussing Mitchell's unpublished paper that argued 75 percent of the cereal price increase was due to biofuels[3] (where low stocks and developing country policy responses were attributed to biofuels), we argued there is no reason to focus on quantities, stock levels, and developing country policy responses if, instead, one analyzes the price links as we do. Recognizing the impact of the price links, one would have enough to explain all of the corn price increase, and much more (after all, the change in corn price due to the biofuel-policy-induced ethanol price premium was greater than the corn price itself!)

So the question becomes, "what is going on?" Put simply (and explained in more detail throughout this book), there were excess profits in ethanol production (profits above levels expected with normal competitive markets) because of capacity constraints Additionally, the intercept of the ethanol supply curve was above the free market ethanol price so there was a redundancy in the ethanol price premium over free market levels.[4] This occurred right through 2006 and 2007 but, in 2008, corn (and soybean) prices were locked onto crude oil prices all the way up to the peak and all the way down, all because of biofuel policies. In no way could other factors have affected the corn price. In other words, regardless of supply/demand shifts in grains and oilseed markets, corn (and soybean and, therefore, all oilseed and coarse grains) prices were determined by crude oil prices all because of this new price link (through biofuel and gasoline/diesel) to crude oil prices.[5]

Thus, we urged Mitchell to adopt the price-link approach to explain the price movement occurring in 2008 (stocks, bad weather, and developing country policy responses could not have had a major impact). It would, therefore, be far easier to explain and convince policymakers and the profession that biofuels were the main reason for the price increases at the time. Mitchell's script was already written and, therefore, not changed. The only thing we agreed on was that biofuel policies had a very big impact on crop prices. Mitchell's paper was ridiculed by the profession for being too simplistic and for attributing a far too high a share of the food commodity price increase to biofuels. However, for reasons just given, we think close to 100 percent of the price increase in 2008 would have occurred in any case due to biofuel policies. Our empirical results later in the book shows that, for a four-year time period, approximately 80 percent of the price increase would have occurred regardless of other market developments [so the estimate in Mitchell (2008) of biofuels direct impact was, if anything, substantially underestimated].

Our theory of how the newly established crop–biofuel–energy price links just described impacted food commodity crops is based on the premise that there is a *law of one relative international price* of grains and oilseeds because of substitution in demand and competition for land. Therefore, their prices follow each other. Roberts and Schlenker (2009) show that the prices of calories are highly correlated and Wright (2014) adopts the same approach by analyzing the quantity of total calories. Therefore, for example, what happens to corn markets because of biofuel policies reverberates in the wheat market and vice versa.

Biofuel policies—sourced in environmental, energy, agricultural, and international trade legislation—are critical. There would be no crop–biofuel price links without them, and no impact of biofuels at least through 2008. There are now two states of nature for biofuel prices: they are linked to energy (gasoline/diesel and crude oil) prices; or biofuel prices float up and away from energy prices. Barring capacity constraints and abnormal profits in ethanol production, crop prices are linked to biofuel prices regardless of the state of nature and, therefore, the new counterfactual is the crop prices evolving according to the crop–biofuel price link, not the good old days when normal supply/demand factors ruled food commodity prices.

Biofuel policies in some circumstances also delink biofuel (corn) and energy (gasoline/diesel and crude oil) prices, thus creating two states of nature. Biofuel (corn) and energy prices are locked onto each other when the tax credit is binding; otherwise, there are *mandate premiums* where biofuel (corn) prices float up and away from energy prices. Why are these two states of nature important? If corn prices are locked onto crude oil prices through the tax credit, this provides a floor for corn prices, and supply/demand shocks in corn markets have no impact on corn prices (unless a *change* in ethanol production affects world crude oil prices—a modest effect at best). Changes in input costs, inventory levels, Asian demand, technological change, and so on do not matter in this case. We show later in the book that 80 percent of the price increase would have occurred regardless of whatever else happened. Because this was the lowest that prices could go. Only when biofuel prices float up and away from energy prices can crop supply/demand shocks affect corn prices (but only by the amount of the mandate premium above the tax credit amount).

Therefore, we could never understand why economists made such a big deal about the estimate that 22 percent of the price increase was due to biofuels (e.g., Roberts and Schlenker 2013) or international nongovernmental organizations (NGOs) like Oxfam (2008) acting like it was a catastrophe by claiming 30 percent of the price increase is due to biofuels when we show at least 80 percent of the price increase would have occurred anyways because of the biofuels' price link to crops, regardless of stock levels or developing country policy response (so even Mitchell's 2008 share of the price increase due to biofuels, adjusted for low stocks and developing country policy response, was far too low!) In fact, we will show that there was (another) $1 billion per day "farm subsidy" in the form of transfers from crop consumers to producers due to biofuel policies (independent of the famous $1 billion per day farm subsidy as calculated by the Organization for Economic Co-operation and Development (OECD 2013).

This has been a story that the world has not seen or understood for far too long and, therefore, the purpose of this book is to rectify that. Ever since our 2008 World Bank meeting, a book of this nature was on our minds to convince policymakers and the profession that this was the proper way to look at biofuel policies impact on food commodity prices.

Returning to the shot that was not heard around the world in the six months leading up to February 2007, although it was a big bang that economists should have focused on, we believe that it was nowhere near as big as it could have been. If there were no production capacity constraints in 2006 and 2007 (and so no excess profits) and the intercept of the ethanol supply curve was not higher than the free market ethanol price, then with a corn–ethanol price multiplier of approximately four, there would have been a much bigger price increase than had occurred. Therefore, rather than being a silent tsunami catching everybody by surprise in the last half of 2007 and first half of 2008, it really was a loud earthquake in the six months leading into early 2007 that everybody missed but could have been louder if there were no abnormal profits in ethanol production or redundancy in the ethanol price premium!

We acknowledge that this is a very different perspective on the entire issue of the role of biofuel policies in affecting grain/oilseed prices compared to what the reader has seen anywhere else. Therefore, this book is controversial. The price boom since late 2006 has been unusual and new tools are required to truly understand the source and cause of the volatility. However, the contents of this book are only the beginning; much more work has to be done.

Other economists generally argue that the big price boom has been caused by traditional supply and demand shocks. Yields for major crops have been growing slowly for decades—comeuppance for underinvestment in public R&D for agriculture—whereas food demand had been increasing, especially in Asia. Economists generally agreed high prices reflected supply-and-demand fundamentals, with growth in demand outstripping that of supply for several years, resulting in lower stock levels. Economists stuck to this old view to explain the economics of grain market behavior, namely the Marshallian supply/demand model and intertemporal price smoothing via storage. They cannot be blamed for this. For the most part, those who work in food economics have not had any interaction with biofuel economics or policy and simply had not considered how the market may have transformed. They believe(d) that because of inelastic short run supply and demand curves, food commodities in world markets are, therefore, susceptible to large shocks, including government policy responses, rising input costs due to energy prices, exchange rate volatility and other macroeconomic factors, and adverse weather shocks. Economists were quick to point out that there was an overabundance of factors that caused this silent, unexpected price boom; it was dubbed the "perfect storm," a conjugation of seemingly unrelated factors that generated this price shock. But we will ask how long does a storm last? And perfection?

We reject this theory and all others as well, including those that focus mostly on inventory adjustments, numerical models that generate "bubble"-like price "explosions" with "mean reversion," combined with sectoral supply/demand shocks, to explain the price levels and volatility. How biofuels impacted the market was the latter's only surprise (getting the better of most economists). The market, for example, knew and knows China's gross domestic product (GDP) growth was over 9 percent per year since 1979, and that Midwest flooding is a common occurrence. Many economists argue one should not and cannot apportion the role of biofuel policy on the price rise and volatility. Others believe biofuels are just a fad or only had a "ripple effect" whereas still others argue the markets needed to adjust[6] or that their analysis is "balanced and comprehensive" when including a multitude of factors to explain the price boom (Headey and Fan 2010).

The fallout of these price surges was tremendous as urban riots and political unrest erupted in over 40 countries; some governments were toppled and political instability became an international concern for overall security (Barrett 2013). Foreign investors sparked a new global land rush, sparking political concerns. The world has experienced a prolonged period of price volatility in global food commodity markets. Historically, these price booms are rare and occur for a short time; therefore, the world's attention is temporarily galvanized. However, this time the world is waiting for when it is going to end. It may not end for a long time, because this time is different. The primary cause was biofuel policies and the crop–biofuel price links that were created.

Many international organizations argued that shielding food security against the threat of more frequent bouts of turmoil in global food markets must be put at the top of the political and economic agenda (Prakash 2011). The need for a new policy dialog was recognized, yet it provoked controversy (The Interagency Report to the G-20 2011). Many international organizations and NGOs believe that improving food security is one of the most pressing issues of our time. The Food and Agriculture Organization's report in HLPE (2013) shows that biofuels are at the center of that debate. The outcome is uncertain. Perhaps one of the most important implications of biofuel policies is the irreversible nature of biofuel policies: the impact of removing all these policies now is very different than not having the policies in the first place. One cannot easily eliminate food–biofuel competition because it is irreversible *ex post*. This leaves hundreds of millions of people vulnerable to price shocks. Safeguarding food security in volatile global markets is a complex task, given the reasons prices behave as they do in this new biofuels era. The future of humankind will depend on a secure and safe availability of food, and biofuels have forced upon us new competition for food.

Food systems successes in the past 70 years brought dramatic poverty reduction and improved standards of living. Approximately five billion people have adequate calories and macro/micronutrients today, up from approximately two billion 50 years ago. Because of public and private agricultural research and policy reforms in developing countries, agricultural productivity growth has outpaced demand growth. The increased efficiency of input use led to lowering real food prices through the mid-2000s. This progress and the income growth associated with it lifted hundreds of millions from poverty, enabled population growth and urbanization over the "Long Peace" of the late twentieth century (Barrett 2013). But…along came biofuel policies. Barrett (2013) notes global population will demand 70–100 percent more food by 2050 than the world consumes today because of a more populous, urban, and wealthier world; this has major implications for commodity composition of diets and trade. There is limited capacity to reduce consumption by reducing food waste, overconsumption in diets, and rebalance diets away from animal products (Barrett 2013).

Therefore, a careful analysis of biofuel policies and their impact on prices is in order as they lie at the heart of the most pressing societal issues facing agriculture. The global food price crises of 2007–2008 and again in 2010–2011 and again in 2012–2013 underscore the mounting challenge of meeting increasing food demand in the face of increasingly scarce natural resources. As a result, poverty and hunger have become high-level policy concerns.

The specter of widespread food insecurity and sociopolitical instability provides the policy background for this book, which simply explains why prices are high and volatile. How to achieve food security in volatile global markets is not our goal, nor will we recommend policies. This book discusses only the causes, not the consequences or policy implications of the recent surge in international food prices although our framework of analysis has implications for the latter; it is just that this book does not get into that (except for some hints here and there).

This book will not evaluate whether the price boom due to biofuels was good for world food security and poverty rates. Some economists argue poverty is more prevalent in rural areas, therefore, poverty rates declined. Others say poverty rates continued to decline regardless of biofuels. Yet others argue the world faced food crises in 2008 and again in 2011 that put a lot of stress on poor consumers and that the current price volatility is not good for anybody. We will not get into that debate. Our objective is to make sure the profession understands how it all happened and the key role of biofuel policies in not only precipitating the great price rise but also continually impacting not only price levels but price volatility to this day.

Since late 2006, biofuel policies have the distinction of being the best known but least understood topic in the field of agricultural economics. Although many papers have been written on the topic, we will show how biofuel policies' impact on food commodity prices was sudden and violent, and unanticipated. The new biofuel era has added important implications for policymakers. However, the profession has to first understand what is going on before being able to make an assessment. This is the goal of this book.

ACKNOWLEDGMENTS

This book would not be possible without the support of many friends and colleagues. We especially thank Chris Barrett for inviting us to write this book in the new Palgrave Macmillan series on Agricultural Economics and Food Policy. His leadership and confidence in us were the impetus that pushed us forward. We are forever grateful to Erika Kliauga who was especially helpful in developing an understanding of the inner workings of the economics of the Brazilian sugarcane industry. She helped us make sense of an unbelievably complex and detailed set of data related to the sugar–ethanol–fuel complex in Brazil. Without her sterling efforts, which resulted in developing the model of Brazil's sugarcane ethanol market, this book would be very incomplete. We also thank Justin Shepherd for all of the discussions and his hard work in his Master thesis at Cornell University, upon which Chapter 10 is based. Our special thanks go to Don Mitchell, the arguments with whom crystallized the ideas in this book. We are grateful for many fruitful discussions, debates, and meetings with John Baffes that resulted in invaluable insights.

We thank Gordon Rausser for his stimulating discussions and for inviting us to write a paper with him, in 2012, that emphasized the importance of analyzing short-term price movements by commodity and the role of capacity constraints. We thank Govinda Timilsina for our collaboration in the research on the biodiesel and Brazilian sugarcane models, and the many helpful discussions and insights we received from him. We appreciate Jerry Taylor's support and encouragement to continue researching and writing about the adverse economic effects of biofuel policies. David Zilberman, as always, stimulated a lot of ideas and Jo Swinnen and Julian Alston were very helpful by being provocative on the policy implications of biofuels, thereby giving us extra incentives to make our arguments. Pavel Ciaian was very thoughtful and generous in his discussions of particular segments of this book. We are grateful to Rob Johansson, Seth Meyer, and Wyatt Thompson for the endless discussions on the finer details of the US mandate, which, in the end, became very important and, therefore, we benefited from their foresight. We are eternally grateful to Ian Sheldon for his confidence in us and for his encouragement throughout. Finally, the crew from Brazil, especially Marcos Jank, Andre Nassar, and Luciano Rodrigues, deserves our gratitude for their constant help in understanding the finer details of the policies and economics of Brazil's sugarcane ethanol markets.

Throughout our research on the economics of biofuel policies, we have benefited greatly from arguments and comments made and information and suggestions received from colleagues and friends, especially Phil Abbott, Kym Anderson, Bruce Babcock, David Blandford, Elizabeth Byrd, Brian Dillon, Koos Gardebroek, John Groenewegen, Tom Hertel, Gal Hochman, Stan Johnson, Madhu Khanna, Harvey Lapan, GianCarlo Moschini, Will Martin, Harriet Mugera, Carl Pray, Deepak Rajagopal, Ford Runge, Josef Schmidhuber, Todd Schmidt, Yves Surry, Peter Timmer, Simla Tokgoz, Bill Tomek, Wally Tyner, Yacov Tsur, and Brian Wright.

Finally, we would like to thank Sarah Lawrence, our Palgrave Macmillan editor, for all her efforts and patience in getting the book in the production stage.

We gratefully acknowledge the financial support received for the research done in this book from: (i) the National Research Initiative of the National Institute of Food and Agriculture, US Department of Agriculture (USDA; Washington, DC, USA; Grant Nos. 121–8520 and 121–564 under the Agricultural and Food Research Initiative (AFRI)); (ii) Cornell University's Agricultural Experiment Station federal formula funds (Project Nos. NYC-121–7433 and NYC-121–7438) received from Cooperative State Research, Education, and Extension Service, USDA; (iii) the "Knowledge for Change" trust fund of the World Bank; and (iv) the European Community (under Project no. 26220220180: Building Research Centre "AgroBioTech." As always, any opinions, findings, conclusions, or recommendations expressed in this publication are those of the authors and do not necessarily reflect the view of the USDA or the European Commission.

ABBREVIATIONS

DDGS	Dried Distiller Grains with Solubles; a co-product of ethanol production
EIA	The Energy Information Administration
EPA	The Environmental Protection Agency
EU	European Union
FAO	The Food and Agriculture Organization of the United Nations
FAO HLPE	The FAO High Level Panel of Experts on Food Security and Nutrition
FAOSTAT	The statistical database of the FAO
GARCH	Generalized AutoRegressive Conditional Heteroskedasticity
GDP	Gross Domestic Product
GEEG	Gallons of Energy Equivalent Gasoline
GHGs	Greenhouse Gases
HFT	High-Frequency Time series
IMF	International Monetary Fund
KELG	Kilometers Equivalent Liters of Gasoline
MEGD	Miles Equivalent Gallons of Diesel
MEGG	Miles Equivalent Gallons of Gasoline
MTBE	Methyl Tert-Butyl Ether; a competitor to ethanol as a fuel additive
NGO	Non-Governmental Organization
OECD	The Organization for Economic Cooperation and Development
PSE	Producer Support Estimate for agriculture as measured by the OECD
RFS	Renewable Fuel Standard
RIN	Renewable Identification Number
RVO	Renewable Volume Obligation

UN	The United Nations
US	The United States
USDA WASDE	The US Department of Agriculture World Agricultural Supply and Demand Estimates
WTO	World Trade Organization

INTRODUCTION

Biofuel policies remain one of the most misunderstood and controversial set of government actions that has ever impacted contemporary food commodity price and policy analysis. The goal of this book is to rectify this by carefully explaining how it all happened. There have been eight years of turmoil in world food grain/oilseed prices. Biofuel policies effectively began a new (modern) era for agriculture, particularly affecting grains/oilseed prices directly. It all started in late 2006 with the six-month boom in corn prices and by the second half of 2007, all food commodity prices caught fire, tripling from their 2005 levels, followed by a collapse in prices in the second-half of 2008 and early 2009. Economists quickly pointed out that this was just another boom and bust as in previous decades, with sectoral supply/demand shocks, inventory adjustments, and macro-economic factors explaining pretty much everything. However, then food commodity prices marched straight back up in 2010–2011 and were still high in 2012. Even in 2013, with record world grain supplies and stocks, prices were relatively high with corn prices double their historical levels prior to late 2006 and commentators were puzzled by the entire grain–oilseed complex into 2014, with exclamations like "beans are in the teens," how can that be? Well, because this time it is different. And the difference is simply biofuel policies.

The nature and causes of the recent commodity price booms and price volatility have been the subject of a rancorous and international debate. Sen (1998, 144) warned against the appeal of

> "Instant economics"…Of course the need for speed is genuinely important in matters of food…But instant economics is also highly deceptive and especially dangerous in this field. Millions of lives depend on the adequacy of policy response to the terrible problems of hunger and starvation in the modern world.

Sen (1998, 144) continues:

> Past mistakes have been responsible for the death of many millions of people and the suffering of hundreds of millions, and this is not a subject in which short-cuts in economic reasoning can be taken to be fairly costless.

The standard approach in this (young) literature has been to dust off and apply age-old methods like standard Marshallian supply/demand curves and the competitive theory of storage. Supply and demand shocks are central to each approach but, with stockholding behavior, the impact of a shock on prices is highly non-linear (in other words, the price change is higher when stocks have been depleted).[1] Emerging from this business as usual, the instant economics approach to grain and oilseed price determination is the standard spiel of modeling biofuels as a shift in demand for crops, causing prices to rise, but, of course, the quantity of biofuels is not that large in the scheme of things, taking up at most a couple of percentage points of total world cropland area.

Therefore, economists had to take shortcuts to explain why, for example, rice prices doubled in a matter of months in 2008. The arguments morph into the huge price impacts of developing country policy responses, unprecedented buying pressure from new financial index investors in commodity futures markets, and numerical models simulating stockholding behavior that generate "explosive" price run-ups.[2] Many other economists argued it was a "perfect storm" of coincidental factors. The constant repetition of keywords reveals all: it is "complex"; it was a boom and bust (like always); mean reversion (to the previous trend of declining real prices); supply/demand shocks like a drought in Australia; low stocks means prices are very sensitive to market shocks, like fires in Russia; and the new financialization of commodity markets and speculation. In each case, the *quantity* of biofuels and supply/demand shocks are the focus of attention and the source of food commodity price changes.

While *The Economist* (2008a, b) described it was a "silent tsunami" and caught everybody by surprise, we will argue it was not silent at all and, therefore, should not have been a surprise if one looks at it from the vantage point of the crop–biofuel price links that were created beginning in September 2006. It is because of these *price linkages* that biofuel policies have a dramatic and unique effect on food commodity prices. Instead of quantities, the price links should be the focus of attention and the specific complexities of the myriad of biofuel policy instrumentalities employed and a plethora of interaction effects between biofuel policies themselves within or across countries, between biofuel policies and other policies (in particular, farm subsidies and developing country policy responses to the high prices), between biofuels, and between commodities.[3] Therefore, the economics of commodity–price volatility have become more complex with the new link between crop and biofuel prices as biofuel policies expose food commodity prices to new sources of volatility. The volatility now also depends on the source of the shock (oil price versus crop supply/demand shocks) or which biofuel policy determines the domestic biofuel market price (e.g., tax credit versus mandate, or world prices). In fact, short-run events that break the link between crop and biofuel prices also contribute to price volatility. Therefore, a new approach has to be developed because this time it is different.

We argue that biofuel policies made, by far, the greatest contribution to food commodity price levels and volatility. Biofuel production increased sharply, beginning in the mid-2000s. The sudden increase in world biofuel production and the linking of crop to biofuel prices ushered in the modern era of price and policy analysis for food commodity markets. We will show in later chapters how the crop–biofuel price multipliers manifest themselves, and how these links change the entire framework of analysis in analyzing not only food commodity price levels but also price volatility. We are in a whole new era and the task of this book is to put things in perspective and show how it all happened.

There is very little distinction between biofuels and biofuel policies because, through 2008, there would be no biofuels without policies. Presumably after the first high crude oil price boom, there would have been investments by the private sector if environmental regulations and the many other biofuel policies over the years had not existed. Biofuel policies caused a link between crop and biofuel prices beginning in late 2006, resulting in a new counterfactual in which the lowest price that grains/oilseeds can reach is when they are locked onto the oil price through biofuel consumption subsidies. Otherwise, crop prices float up and away from oil prices with mandate premiums. We find that the maximum effect of nonbiofuel factors—crop supply/demand shifts and stockholding behavior—is only 20 percent of the corn price increase in the period 2006–2007 to 2010–2011.

To truly understand the impact of biofuel policies on price levels and volatility, rigorous theoretical and empirical models have been developed for alternative biofuel policies and biofuel feedstocks. This book provides a comprehensive and balanced analysis of the various factors causing the recent turmoil in world grain/oilseed prices, while isolating the impact of biofuel policies. In addressing these controversies, we conclude that biofuels policies were the principal cause of the sudden change in grain/oilseed prices and of the ensuing price volatility. We argue that biofuels policies, under the rubric of environmental and energy policy legislation, were the only surprising event in the food commodity markets; and that the mechanisms of these policies were so complex that it was impossible for market participants to anticipate them. We also maintain that the current discussion should not be centered on whether food grain/oilseed prices are high or low, or what the right price of food should be, but the discussion should rather be about the right food policies.

Our approach is in sharp contrast to the literature where old frameworks of analysis are employed for this new issue of biofuels. The literature emphasizes how the quantity of biofuels impacts food commodity prices and not the overriding effect of the new price links, established because of biofuel policies, among crop (cereals and oilseeds), biofuel (ethanol and biodiesel) and energy (gasoline/diesel), and crude oil prices. This is the focus of our book.

Organization of the Book

This book lays out a unique evidence-based approach to explain the food commodity price boom from September 2006 to 2014. In order to achieve this objective, we develop a clear and comprehensive analytical and empirical foundation to explain price movements that depart from the conventional wisdom and, thereby, explode several longstanding myths. Chapter 1 begins by defining biofuel policies as a strange mixture of policies sourced in environmental, energy, agricultural, and trade legislation. We show how specific US environmental regulations were critical, and that economists are not in agreement in defining biofuel policies. Topics that round out the chapter include data on how world biofuel production has leveled off, the context for the current boom relative to past booms, how prices among crops move together, and specific details for the time period September 2006 to the end of 2007 on how corn and ethanol prices became linked because of biofuel policies.

Chapter 2 is the juggernaut upon which the entire book rests as it provides the core theory of the critical price linkages and economics of blend mandates. Using US corn–ethanol and ethanol–gasoline price links as an example, a methodology is provided to determine the (hypothetical) free market ethanol price, the price link between corn and ethanol (the magnitude and what it depends on), the price links between ethanol and gasoline, and, therefore, the two states of nature: corn prices are lowest when locked onto crude oil (through ethanol and gasoline) prices (with or without tax credits); or ethanol prices float up and away from gasoline prices when "mandate" price premiums appear (due to formal or de facto mandates through environmental regulations, due to situations where consumers have no choice, or prices are determined on world markets). A basic economic theory of blend mandates is developed (with the many implications, compared to when ethanol prices are directly linked to gasoline prices). A rigorous discussion ensues on the empirical results, explaining why the model makes prediction errors for corn prices and, in theory, how low corn prices could have gone in the past.

After first deriving the ethanol supply curve and explaining why there is a gap between the intercept of the ethanol supply curve and free market prices (what we call "water" in the ethanol price premium), Chapter 3 otherwise provides empirical results on the ethanol price premium, and the impact of ethanol policies on corn (and hence other grain/oilseed) prices compared to (i) the new counterfactual where the corn–ethanol price link puts a variable floor under corn prices, and (ii) the old days when no biofuels existed. We determine that corn prices went up 80 percent and 40 percent, respectively, under these two scenarios. We also show there has been, on average, a $1 billion per day "subsidy" to crop producers resulting from high prices due to biofuels policies, and that allowing the US tax credit to expire may save $1 billion per day transfers from consumers to crop farmers alone.

Chapter 4 provides a detailed analysis of the 2007–2008 price boom and, thereafter, focusing on the relative prices month by month for the four main commodity groups: corn, soybeans, wheat, and rice. This chapter provides texture to the theoretical models in preceding chapters and provides a platform to survey the various theories in the literature.

Chapter 5 identifies four general literatures (apart from the one outlined in this book) on how biofuel policies impact grain/oilseed prices: standard sectoral shifts in Marshallian supply–demand curves; the depletion of inventories due to the unexpected shock in crop demand generated by biofuels, rising prices because of the nonlinear relationship between stocks to use ratio and prices; econometric time series regressions with monthly or quarterly data; and a pure statistical approach to analyzing the links among various crop, biofuel, and energy prices using high-frequency data (daily, weekly, or monthly). The rest of the chapter dispels the myths of the "perfect storm" literature that dominate economists' thinking.

Chapter 6 is an exception because it goes into detail on the implications of our theory of biofuel policies for only one of the many "perfect storm" explanations of the food commodity price boom: the impact of developing country policy responses on world grain/oilseed prices in 2008. Contrary to the literature and, therefore, the received wisdom, we find little impact. Detailed analysis of all the other "perfect storm" factors, using our framework, awaits another book, but Chapter 6 highlights the importance of understanding how grain/oilseed prices are determined in this new biofuel era.

Chapter 7 critically assesses the last literature identified earlier, namely that using high-frequency data to frantically seek links between crop (corn, sugarcane, and oilseeds), biofuels (biodiesel and ethanol), energy (gasoline and diesel), and crude oil prices using state-of-the-art time series econometrics techniques. However, this literature ignores the modeling concerns of our theoretical (and empirical) framework, which we will show proves to be fatal as to the relevance of this literature's empirical results—a literature that argues it has policy implications.

Chapter 8 provides a summary of the implications of our theory for price volatility. This chapter also presents an opportunity to survey one other competing literature identified—the one that extends the Marshallian framework to include the demand for storage and the resulting implications on both the shape and placement of the total (consumption plus stock) demand curve. This literature, like the Marshallian supply/demand model, is still relevant in this new era but only to the extent biofuel prices "float up and away" from their energy (gasoline/diesel) prices. We also present ideas on how the theory of competitive storage has to be adjusted in this new world of crop–biofuel price links.

Chapter 9 develops the economics of the biodiesel model applicable to both soybeans and rapeseed. Specific focus on European Union (EU) biofuel policies is given as EU biodiesel is the predominate biofuel in both production and consumption, unlike other major biofuel-producing countries. Unlike the corn–ethanol with a *co-product* returned to the market as feed (in Chapter 2), the soybean/canola oil–biodiesel model differs because oilseed meal and oilseed oil extracted from the crushing of the oilseed are *joint products*. Analogous to corn–ethanol, we provide a soybean/rapeseed oil–biodiesel price prediction equation and describe its performance. The chapter also describes some of the EU trade policy effects on biofuel prices and trade, and analyzes various aspects to determine the (in)efficiency of biodiesel versus corn–ethanol.

Chapter 10 details the complex nested structure of the US biofuel mandate and the interactions between the sub-mandates leading to implications for the ethanol "blend wall" (where regular cars can only absorb a maximum of 10 percent ethanol). There are several important findings: the US ethanol blend wall has been "binding" since 2010 and has been overcome by biodiesel, not significant expansion in E85 sales; the complex structure of the renewable fuel standard (RFS) is responsible for emerging world biofuel trade patterns (the United States has become a net exporter of ethanol and a major importer of biodiesel) and relative prices between ethanol and biodiesel; and the nested structure of the US mandate explains capacity constraints and, therefore, the prediction errors in our soybean oil–biodiesel price linkage equation. This chapter clearly explains, contrary to conventional wisdom, that the US mandate is implemented as a blend mandate, and how renewable identification numbers (RINs) prices are determined, and the implication of their convergence among biofuel categories.

Chapter 11 develops a model of the sugarcane–ethanol complex in Brazil, which differs from corn–ethanol and soybean oil–biodiesel because sugar and ethanol are *competing products*, there are two *by-products*: molasses (from sugar production) used to make ethanol and bagasse from all sugarcane process used for electricity, and two consumer demand curves: one for a fuel mixture of 25-percent ethanol (E25, like E10 in the United States) and a well-functioning and competitive E100 market (unlike that for E85 in the United States). The Brazilian model rounds out the theory of the three main feedstocks for biofuels: corn, oilseeds, and sugarcane. The chapter analyzes the impacts of the blend mandate, gasoline tax, holding gasoline prices below world prices, and differential tax exemptions between hydrous and anhydrous ethanol. These controversies plague Brazilian politics and, because of trade in ethanol, impact US ethanol and, therefore, corn prices as well.

Chapter 12 highlights the various interaction effects between biofuel policies, between biofuel and farm policies, and between commodities and countries. International trade is shown to have a differential impact on such interactions. Biofuel policies are also shown to have caused new political interactions for the US farm bill and technological interactions where irreversible investments have important policy implications.

The final chapter offers some concluding remarks by putting our analysis in perspective, and outlining a way forward as to research priorities. We also speculate as to possible situations where the model we develop in this book breaks down as to its impact on grain/oilseed prices and, therefore, has potential implications for price analysis in the future.

CHAPTER 1

HOW BIOFUEL POLICIES USHERED IN THE NEW ERA OF HIGH AND VOLATILE GRAIN AND OILSEED PRICES

So far, we have emphasized the importance of the crop–biofuel–energy–crude oil price links, and the resulting two states of nature where corn prices can go no lower than when locked onto crude oil prices through ethanol and gasoline prices, and with tax credits.[1] To illustrate the importance of price links and how quantity (sectoral supply/demand) shocks may not be the principal driving force in explaining food commodity price levels, consider Figure 1.1. Production of biofuels accelerated in mid-2000s and has now leveled off. Many commentators use that fact along with the pronounced volatility in grain prices in the meantime as a proof that biofuels are not the leading cause of high prices (e.g., Hamelinck 2013).

But ironically, in the time periods when grain prices decline in Figure 1.1, mandate premiums (ethanol price premiums above and beyond the tax credit) kick in so the relative impact of biofuel policies on corn prices is even higher. Another typical example is Baffes and Haniotis (2010, 12) who argue:

> Yet, worldwide, biofuels account for only about 1.5 percent of the area under grains/oilseeds...This raises serious doubts about claims that biofuels account for a big shift in global demand.

A small share of total output means a small impact on price; this is the conventional wisdom. Baffes and Haniotis (2010, 12) go on:

> It is striking that maize prices hardly moved during the first period of increase in US ethanol production, and oilseed prices dropped when the EU increased impressively its use of biodiesel.

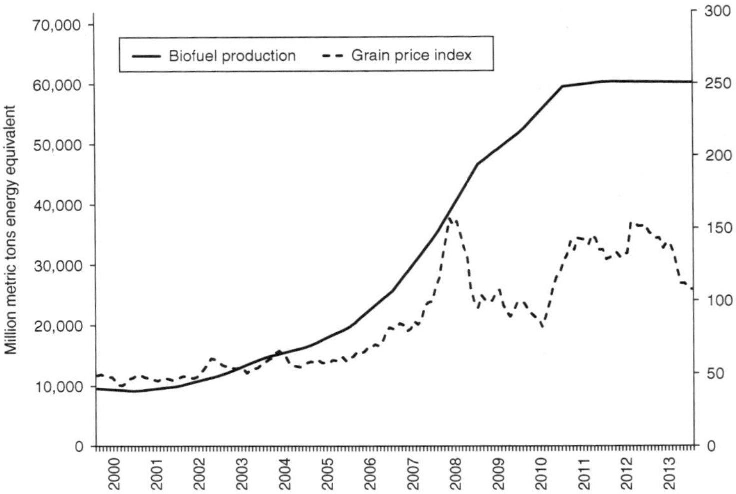

Figure 1.1 Grain prices versus biofuel production.

Source: British Petroleum (2013), World Bank (2014).

However, our interpretation of Figure 1.1 is quite different than that of the standard literature: so much of the impact of biofuel policies is summarized in price links; there is a lot less room for quantity shocks to matter in this new era. With these price linkages, in theory, the quantity of crops dedicated to biofuels can be irrelevant. Searchinger (2013) argues that the higher the price increase due to biofuels, the lower the supply response and, therefore, the lower the indirect land use change. However, such an analysis relies on traditional supply/demand analysis. Our analysis shows that it is very possible to have high price and land-use-change impacts simultaneously (or vice versa); again, this exemplifies the implications of our basic framework developed in this book that quantities may not matter as they did before the crop–biofuel price links were developed.

Before setting out to show how this crop–biofuel price link was formed, we first have to define biofuels and biofuel policies.

Defining Biofuels and Biofuel Policies

We first have to define what a biofuel is (see Box 1.1). For the purposes of this book, it is ethanol and biodiesel blended with gasoline and diesel, respectively, and used solely as transportation fuels.

Box 1.1 What is a biofuel?

Biofuels analyzed in this book are liquid fuels produced via various chemical processes from biological sources (plants or animals or "biomass"). Biofuels can be solid, liquid, or gaseous, but we focus on liquids that are substitutes to conventional fossil fuels (i.e., gasoline and diesel) and can be consumed either in their pure form or—most often—in blends with conventional fossil fuels. The biomass for the production of a biofuel is referred to as a biofuel feedstock. Currently, the bulk of ethanol (alcohol) is produced from agricultural crops such as corn, wheat, sugarcane, sugar beet, cassava, barley, and sorghum. Biodiesel (fatty acid ester) is made mostly from food crops such as soybean, rapeseed, and palm oil but also from recycled vegetable oils and animal fats.

Liquid biofuels are often classified into first-, second-, and third-generation biofuels. Biofuels from food crops and animal oil fats (those mentioned earlier) are first-generation whereas second-generation biofuels refer to ethanol from lignocellulose and organic wastes (e.g., miscanthus grass, corn-stover, tree material, waste wood, and paper); third-generation biofuels are from hydrogen produced either by gasification of lignocellulose or directly from microalgae. A fourth-generation biofuel would be one that is carbon negative (possible through genetic engineering of crops).

We focus on two main kinds of biofuels: ethanol and biodiesel. (As we discuss in Chapter 10, "renewable diesel" is made by a separate process than "biodiesel," but for the purpose of this book, we lump them together). Ethanol is blended with gasoline and biodiesel is blended with diesel. Each feedstock typically produces only one type of biofuel, but some technologies have been developed to manufacture biodiesel from an ethanol feedstock—for example, corn oil extracted from the co-product of corn processed into ethanol is used for biodiesel production.

Common blends of ethanol include E10, E15, E85, and E100 (in Brazil) with 10, 15, 85, and 100 percent of ethanol in the final fuel mix, respectively. For biodiesel, these blends are B2, B5, and B20 (2, 5, and 20 percent of biodiesel).

One will come across a good number of papers describing how ethanol was first used at the turn of the last century in Henry Ford's Model-T car but we argue throughout the book that this time is different (in terms of ethanol's impact on agricultural markets) and, therefore, the historical use of ethanol is not relevant.

The US Energy Information Administration (EIA) (2013) projects biofuels will grow from 2 percent of transportation fuels to 4 percent (energy equivalent) in 2035 under current policies. Therefore, the current use of biofuels in total transportation fuels is very small and is to double in 20 years (but from a small base). Organization for Economic Cooperation and Development (OECD)/Food and Agriculture Organization of the United Nations (FAO) (2014) forecasts that biodiesel's share of 25 percent in 2013 will stay the same in 2020. We will develop measures to assess the relative efficiency of biodiesel and ethanol in terms of price per mile achieved by consumers and their relative impacts on crop prices; we conclude that biodiesel is not nearly as inefficient as other economists claim, but the issue requires further research. International trade is a small share of production but is growing, and it is enough to link world biofuel prices; we will show in later chapters that this has important implications for crop price levels and volatility.

Although biofuel production and consumption are mostly concentrated in the United States, Europe, and Brazil, more than 60 countries have implemented biofuel policies. The other more important countries for biofuel production include Argentina, Canada, China, and India.

Next, we have to define what a biofuel policy is. Biofuel policies are a strange mix of rules, regulations, and market interventions that are sourced in environmental, energy, agricultural, and trade legislation. The policy toolkit is quite diverse, complex, and ever evolving and not always intended to impact biofuels and crop prices—the consequences were often unintended and neither understood nor anticipated. The main subsidy programs supporting the biofuels industry are twofold. First, market prices for biofuels are supported through blend mandates (a minimum percentage of total fuel requires biofuels; the subsidy conferred to biofuel producers from blend mandates that provide a guaranteed market for their product above free market levels). Second, biofuel consumption subsidies that by themselves raise the market price of a biofuel (tax credits in the United States and full or partial tax exemptions in all other countries that are fully applied to competing products—gasoline and diesel fuels).

The secondary yet important categories of biofuel policies include production subsidies, for both biofuels and feedstocks (e.g., for corn), including research and development (R&D) grants (promoting the development of biofuel projects or technologies), provided incentives for ethanol producers in the form of insured loans, tax write-offs, investment subsidies, price guarantees, and purchase agreements, and many federal and state "infrastructure" subsidies were created, such as subsidies for alternative vehicles and fueling stations. Export subsidies and import barriers—along with binary sustainability standards where biofuels from different feedstocks have different greenhouse gas (GHG) emission reductions relative to the fossil fuel they are assumed to replace—are also important ongoing policies. Relative excise fuel tax rates are also important (e.g., the European Union has diesel subsidies and both the European Union and the United States have higher tax exemptions for biodiesel). Perhaps the most important biofuel policies have been the environmental regulations that have impinged on requiring the increasing use of ethanol over other fuel additives (see Box 1.2).

Box 1.2 Do economists agree on what a biofuel policy is?

Although a thousand papers or more have been written on the economics of biofuel policies, surprisingly, economists cannot agree on what a biofuel policy is. Biofuel policies are different things to different people but, in reality, are a strange mix of rules, regulations, and market interventions that are sourced in environmental, energy, agricultural, and trade legislation. The policy toolkit is quite diverse and includes mandates, tax exemptions, subsidies for biofuels and their feedstocks, tax write-offs, investment subsidies, export subsidies, and import barriers. Moreover, there are binary sustainability standards where biofuels from different feedstocks have different GHG emission reductions relative to the fossil fuel they are assumed to replace. Policies are enacted at all levels of governments and have complex interaction effects, domestically and internationally (as it will be obvious as one reads through the book but see especially Chapter 12 for a more formal analysis of a few specific policy interaction effects).

As we have shown in this book so far, one has to recognize that the key influence of biofuel policies was to cause the link between crop and biofuel prices. The two key events for the corn–ethanol price link were the activation of the tax credit with high crude oil prices and the ban on methyl tertiary butyl ether (MTBE), a lower cost fuel additive that competes with ethanol. To understand the nuances of "biofuel policies," just look at the US Clean Air Act regulations alone. These regulations can be construed as the key "environmental" policies leading to the crop–biofuel price links and the food commodity boom in 2007–2008 and thereafter.

The relevance of the Clean Air Act for food commodity prices began a long time ago with the ban on leaded gasoline that led to the adoption of MTBE. Then in the 1990s, reformulated gasoline (RFG) was mandated by Congress where more oxygenates like MTBE and ethanol were required to reduce local air pollution. Earlier, there was the 1984 federal legislation on leaking underground storage tanks (LUSTs) for gasoline that was not strictly enforced. LUSTs led to state bans on MTBE in the beginning of the 2000s. This was followed by the federal Renewable Fuel Standard in 2005 (to replace the RFG), the 2006 federal court decision not to grant immunity to lawsuits on using MTBE because of LUSTs (ethanol prices then reached their all-time peaks), the doubling of the original mandate legislated in 2005 in December 2007 (the current RFS), and, finally, the irreversible investments (due to the RFS) in gasoline production to use 10 percent ethanol as an octane enhancer (which has implications for the effects of removing a policy); see Chapter 12 for the effects of irreversible investments.

So what is the controversy? Some economists do not think any of these environmental regulations are public policies. According to Babcock (2013), only the 2005 and 2007 mandates are policies and all of the other regulations are "market forces." The Federal Reserve Bank of Kansas City (Kauffman 2012) correctly interprets Babcock's views with a headline blaring "Markets, Not Mandates, Shape Ethanol Production"! But we argue that each of these Clean Air Act rules, regulations, and mandates is a biofuel "policy"; from one end (a ban on leaded gasoline) to the other end (where the RFS caused refiners to make irreversible investments). One should not pick and choose what a biofuel policy is and what it is not, just to conveniently make a point.

It is important to recognize that the first two categories of biofuel policies (in other words, tax credits/exemptions and mandates) do not, by themselves, discriminate against international trade. However, the other policies listed earlier do. For a good summary of all of these policies, see Chapter 1 in the High Level Panel of Experts on Food Security and Nutrition (HLPE) (2013), the Nuffield Council on Bioethics (2011), and FAO (2008b).

Biofuel policies have arisen as a confluence of policies that are a subset of energy, environmental, agricultural, and trade legislations designed to obtain multiple goals, including energy security, environmental quality (reduced local air pollution and GHG emissions), improved farm income and rural development, and induced technological change. Other stated political goals include increased domestic energy production that creates "green" jobs and foreign exchange savings with reduced crude oil imports.

Policies are enacted at all levels of governments and have complex interaction effects, domestically and among countries (as will be shown throughout the rest of this book).

An important aspect of biofuel policies, apart from creating the crop–biofuel price links with high multipliers, is the irreversible nature of biofuel policies: the impact of removing all of these policies now is very different from not having the policies in the first place. This has added important implications for policymakers. Our emphasis in this book is on the impact of implementing these policies in the first place.

The Recent Food Commodity Price Boom in Perspective

The first thing to note is the broad and sharp co-movement among crop prices (see Figure 1.2).

This reflects the consumption substitution possibilities among crops and that they compete for the same land (see Box 1.3).

As shown in Figure 1.3, following three decades of decline, international food grain and oilseed prices have spiked three times in eight years. Note that crude oil prices led agricultural prices in the most recent boom period, unlike in the first boom of the late 1940s when these two prices

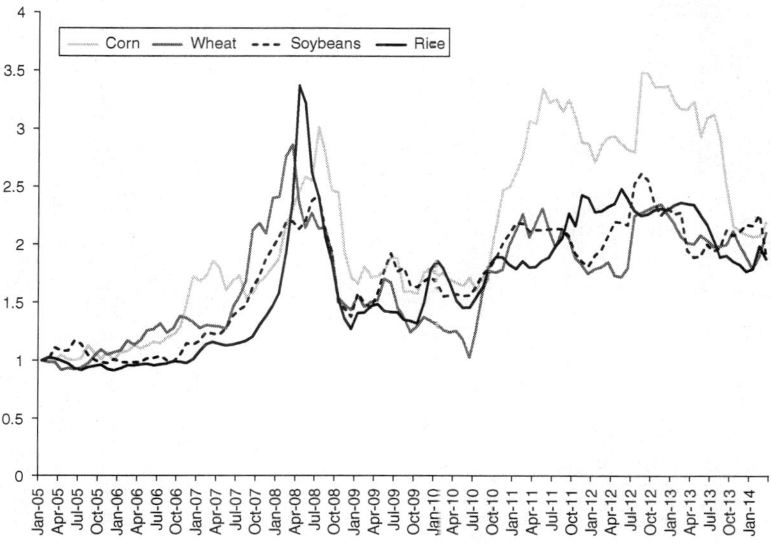

Figure 1.2 Cereal and oilseed price indices (January 2005 = 1).

Source: World Bank (2014).

Box 1.3 The law of one relative international price for grains and oilseeds

Figure 1.2 shows sharp co-movement of food grain/oilseed prices. This is because of competition for the same land (on a global basis) and substitution in demand (e.g., up to 25 percent of world wheat production has been consumed as feed for livestock). de Gorter et al. (2013a) provide empirical evidence on how these prices move together. Roberts and Schlenker (2013) aggregate the four main commodity groups into a single commodity, weighted by calories, noting that the price of calories for all grains taken together is highly correlated with the price of each grain in the past (with a correlation coefficient of at least 0.93) (Roberts and Schlenker 2009).

We call this "*the law of one relative international price of grains and oilseeds*" and this "law" is critical when we study the impacts of biofuel policies on prices throughout this book. Indeed, we will show that biofuel policies integrate grains and oilseed markets even more (e.g., corn oil extracted from the ethanol's co-product is higher in protein and its price is positively correlated with the wheat price). The data in the table below show how relative prices have not changed since September 2006.

Table B1.3.1

	Crop price over corn price		
	Wheat	*Soybeans*	*Rice*
January 1964 to March 2014	1.3	2.3	2.6
January 1964 to September 2006	1.3	2.3	2.6
October 2006 to March 2014	1.3	2.3	2.4

Source: World Bank Pink Sheets.

were unrelated or in the 1970s when agricultural prices led crude oil prices. Is this a hint that this time is different? Did crude oil prices lead food commodity prices because of the price links created by biofuel policies? This is the hypothesis of this book. The collapse of prices in 2009–2010 was a special aberration due to the US financial crisis that precipitated the Great Recession, yet agriculture did not drop as much as energy and especially metals. For example, the Illinois farm price of corn plummeted to a low in

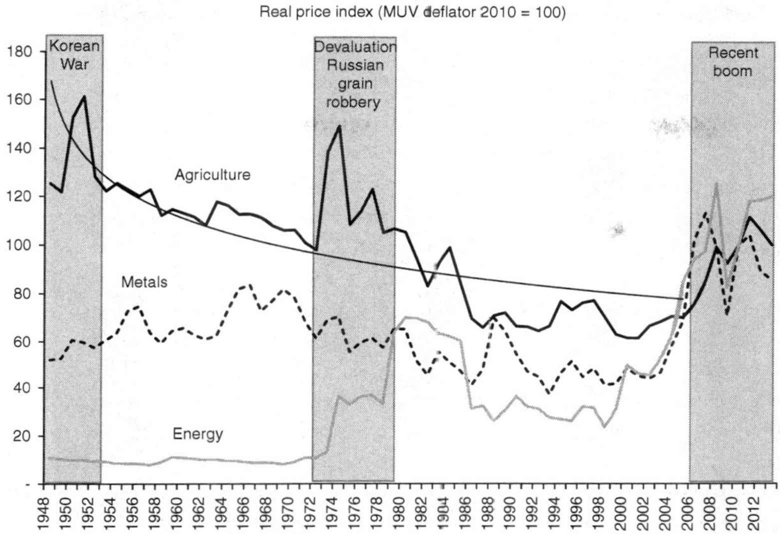

Figure 1.3 Historical price booms and busts in perspective.

Source: World Bank (2014).

2009 that was double the low in 2005–2006 when the world was enjoying its biggest economic boom ever. The co-movement of metals, energy, and agricultural prices is a temptation for economists to attribute the grain/oilseed prices as a macroeconomic phenomenon and, therefore, microeconomic analysis of the like undertaken in this book is not very necessary. We will show that this time is different and macroeconomics play much less of a role this time than before.

Note that the booms have been one-sided: a spike above the long-term moving trend but never below.[2] The issue of price levels versus volatility is less important if one realizes that all food grain price shocks historically are above (none below) the long-term negative price trend for food grains and that, if prices are high, consumers prefer volatility (in other words, for the price to decline) and vice versa for net sellers of grain and oilseed crops. Moreover, as Figure 1.2 shows, because of the law of one relative international price for grains and oilseeds, all crop prices move together; thus, there is very little room for consumers to arbitrage and take advantage of price instability. Nevertheless, prices have been below the long-term trend for two decades: 1985–2005. However, the boom in agricultural prices is longer in the most recent time period relative to previous booms, indicating that the fundamentals of the current boom differ from previous ones.

The Calm Before the Storm: The 32 Months Before September 2006

For 32 months prior to September 2006, Figure 1.4 shows there is no visual link between corn and either crude oil or ethanol prices although ethanol and oil prices more than doubled in that time period. Corn prices were flat at $2 per bushel over the 32 months in which oil prices rose sharply, although corn market participants were well aware corn production is energy intensive and, therefore, the observed corn price was adjusted by rational market participants for the higher input costs (Baffes 2007).

Crude oil prices began their historic rise at the beginning of 2004, breaching the ever-important $40 per barrel mark in mid-2004. Ethanol prices also rose sharply, partly because many individual US states were banning methyl tert-butyl ether (MTBE), a lower-cost but close substitute for ethanol as an additive for fuel consumption, and partly because high oil prices activated the otherwise dormant US tax credit established in 1978, providing a premium on ethanol over gasoline prices. However, corn prices held steady and presumably were adjusted for the indirect link between crude oil and corn prices that always exists through input costs, because corn uses energy-intensive inputs. This factor has been characterized as a major driver of commodity price increases (Baffes and Haniotis 2010;

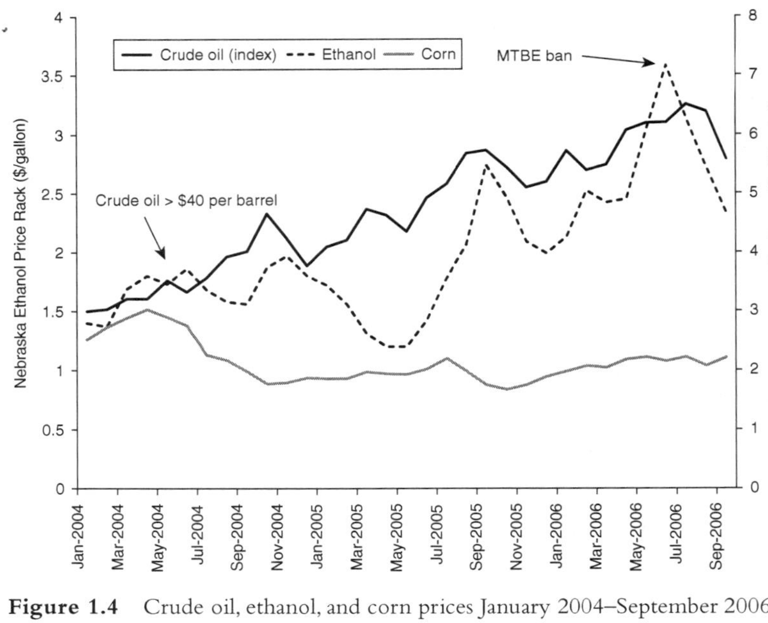

Figure 1.4 Crude oil, ethanol, and corn prices January 2004–September 2006.

Source: Rausser and de Gorter (2012).

Abbott et al. 2008).[3] Ethanol prices soared to a peak of $3.65 per gallon in July of 2006; they never reached that level again, however. Ethanol prices fell toward $2 per gallon for many years thereafter, whereas corn prices peaked at over four times their average level in the first 32 months of 2004–2006 on three different occasions: 2008, 2011, and 2012 (in each year of these years, the corn price peak got ever higher—see Figure 1.2).

The Six-Month Price Boom Not Heard Round the World

Wright (2008, 8) writes

> Crop diversion [for biofuel] can hardly have come as a surprise in 2006 ... usage started no later than 2004 ... clearly foreseeable ... oilseeds for biofuel use ... no surprise.

However, the complacency of corn markets changed suddenly in September 2006 when the Central Illinois farm price of corn rose 88 percent (Kansas City white corn price 107 percent) in six months (September 2006 to February 2007). Figure 1.5 summarizes how, after realizing corn–ethanol processing capacity was to double in one year, markets reacted quite suddenly two-thirds of the way through the year and corn prices began their upward march, causing tortilla prices to double, thereby precipitating the Mexican Tortilla Crisis in January 2007. The ensuing political anxiety led to the February 2007 ban on wheat exports by India and later by Ukraine, the beginning of many developing countries' policy responses to come (Timmer 2008, 2010; Timmer and Dawe 2010; Slayton 2009). Almost all commentators claimed these policy reactions exacerbated grain price increases in 2008 (Anderson et al. 2014a, b). Finally, Figure 1.5 shows the United States doubled its ethanol mandate in December 2007 and, thus, the high grain/oilseed price boom was in full swing. In retrospect, economists should consider the 2006–2007 price boom as the boom before the boom.

By September 2007, corn and ethanol prices became tightly linked (Rausser and de Gorter 2012). By that time, abnormal profits in ethanol production had dissipated. Meanwhile, soybean and rapeseed oil prices were tightly linked to biodiesel prices in the United States and the European Union, respectively. As we will show in later chapters, US exports of ethanol in 2010 caused US ethanol prices to be linked to ethanol prices in Brazil (made from sugarcane) such that shocks in world sugar markets reverberated in the Chicago trading pits for yellow corn. All of this sets the stage for our theory to unfold. We begin to formalize all of this in the next chapter.

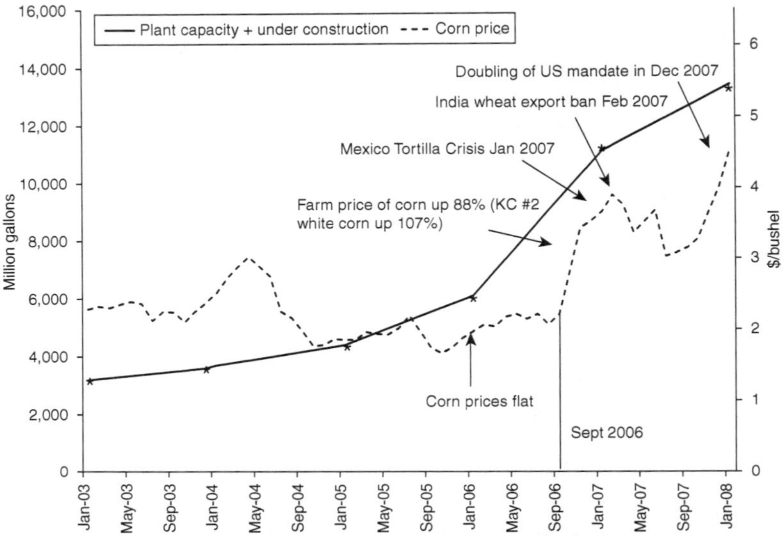

Figure 1.5 Ethanol plant capacity and US farm price of corn.

Source: Rausser and de Gorter (2012).

Concluding Remarks

Grain and oilseed prices exploded unexpectedly at the end of 2006 because of a US environmental policy banning MTBE due to leaking storage tanks, and because of a US energy policy of 1978 that introduced a blender's ethanol tax credit that was subsequently activated by high oil prices in the latter half of the first decade of the twenty-first century. These two events created a direct link between crude oil and corn prices for the first time in late 2006. Later, when the ethanol mandate became binding and created an ethanol price premium greater than the blender's tax credit; biodiesel and non-US ethanol production in the rest of the world also impacted crop prices. Note that biodiesel and non-US ethanol production in OECD countries together make up just half of the US corn–ethanol supply, but more land is used per unit of biofuel produced.

The sudden increase in all grain/oilseed prices due to biofuels policies probably caused the developing countries' policy responses (e.g., export bans) and speculation (e.g., hoarding of rice). It is an open question if these responses caused world prices to increase in 2008 (see Chapter 6 which argues otherwise). Prices accelerated in 2006 and then peaked in 2008. Crop–biofuel prices are now locked onto each other indefinitely, ushering in a new era that is the subject of this book. Biofuel policies also in some circumstances become delinked from crude oil, and hence energy (gasoline/diesel) prices when mandates are binding.[4] Biofuel–energy prices are locked onto each other when the tax credit is binding; if there are mandate premiums, then biofuel prices float up and away from energy prices. Therefore, there are two states of nature: corn and ethanol prices, now locked onto each other for the first time ever, are *lowest* when ethanol prices are *locked* onto gasoline prices which are locked onto crude oil prices. Otherwise, ethanol, and, therefore, corn prices, float up and away from energy prices. *Lowest* when *locked* represents a new counterfactual and is different than supposing no biofuels is the alternative.

These two states of nature are important because, if corn prices are locked onto crude oil prices through the tax credit (if any), then supply/demand shocks in corn markets have no impact on corn prices (except for the effect of the *change* in ethanol production on world crude oil prices, which will be modest at best). Only when biofuel prices float up and away from energy prices can crop supply/demand shocks affect corn prices (but only by the amount of the mandate premium).

Another important aspect of biofuel policies is the strong co-movement in crop prices (see Figure 1.2). All prices adjust quickly to one another because of competition for land on the supply side and substitution in demand. We call this the *law of one relative international price* for grains and oilseeds (see Box 1.3). Witness how prices for yellow corn fed to livestock went up 88 percent in six months in 2006–2007, yet white corn for tortillas in Mexico went up 107 percent—farmers can substitute any grain and oilseed in a very short time period for feed to livestock or for food. The share of wheat fed to livestock went up over 20 percent in the time of the food price crisis (not only more than double the wheat production shortfall due to bad weather in 2007–2008 that so many economists argue precipitated the price boom, but also recurring, unlike the bad weather).[5] Furthermore, approximately two billion people in the world spend at least 50 percent of their incomes on food; therefore, when wheat prices skyrocketed in 2008, consumers had a huge incentive to switch to other commodities (and quickly), bringing these prices up (and stemming the increase in wheat prices).

Of those who recognize commodity prices move together, many argue the co-movement is due to common factors like trends, macroeconomic factors such as exchange rates, world gross domestic product (GDP) growth, speculation, portfolio diversification, and higher input costs for commodities due to crude oil price increases. We agree that there is a correlation between energy, metals, and agricultural prices—see Figure 1.3—and, thus, it is easy to fall into the trap that only macro forces explain food commodity prices. However, it is the purpose of this book to show otherwise by dissecting the different mechanisms biofuel policies impacted grains and oilseed prices, with the law of one relative price being a key input to the story. Our basic working hypothesis is that biofuels was the only market surprise that caused the price shock.

In conclusion, HLPE (2013) asked several penetrating questions, answers to which inform and give the reader of this book a glimpse of what is to come—a "look under the hood." One question was: "*What drives the 'incremental,' separate impact of biofuels on food commodity prices?*" We will show that because biofuel policies caused the price links between crops and biofuels, biofuels drove the rise in food commodity prices—this is not an incremental effect but the bulk of the effect because the new counterfactual is, taking corn prices as an example, always locked onto ethanol prices that can go no lower than gasoline (on a miles equivalent basis). So, the question is upside down; biofuels drove the rise in food commodity prices (through price links). The "up and away" part is the *incremental* impact quantity changes can have on ethanol and, thus, corn prices; not vice versa—biofuels do not have incremental effects—they have the bulk of the infra-marginal or "the no matter what" impact.

Another question HLPE (2013) asked was: "Retrospectively, to what extent did biofuels contribute to the food price spikes and increased level of food prices in the specific context of the last five years (2007–2012)? In that specific context, what was the share of responsibility of biofuels with respect to other factors?" We estimate in Chapter 3 that for the four-year time period of 2007–2008 to 2011–2012, 80 percent of the commodity price increase would have happened regardless of all other factors, given the new counterfactual that corn prices would otherwise be tied to the crude oil price through the tax credit and the ethanol–gasoline price link. And because the law of one relative international price of food grains and oilseeds holds (due to these crops being substitutable in demand and compete for land), a minimum 80 percent of all food grain and oilseed price increases were due to biofuel policies.

A final question asked in HLPE (2013) was: "What could happen in the future? To what extent might biofuel policies contribute to price increases or high prices in the future?" Our analysis later in the book will show that, if crude oil prices were \$150 per barrel today, the corn price would be at minimum \$7 per bushel and if the blender's tax credit were still in place, given recent market parameters, the corn price would be \$9.50 per bushel (and through the law of one relative price, all other crop prices would increase commensurately).[6]

We now turn to formalize our framework.

CHAPTER 2

THE ECONOMICS OF BIOFUEL POLICIES: THE THEORY OF CORN–ETHANOL AND ETHANOL–GASOLINE PRICE LINKS

Wright (2011, 32) argues that:

> Recent price spikes are not as unusual as many discussions imply. Further, the balance between consumption, available supply, and stocks seems to be as relevant for our understanding of these markets as it was decades ago...the tools at hand are capable of explaining the main forces at work.

We, on the other hand, argue that the most important development in grains/oilseed markets since 2006 has been the direct link between crop and biofuel prices, and between biofuel and energy (gasoline/diesel and crude oil) prices. These links represent the first time that these disparate markets have been so closely connected—and the links were initiated by biofuel policies. These new price links, and the complex thicket of policies that undergird it, create the need for new tools to understand and explain the workings of the new marketplace. Thus, we develop a unique comprehensive, integrative framework that carefully analyzes the specific relations between the crop (grains and oilseeds), biofuel (ethanol and biodiesel), energy (gasoline and diesel), and crude oil prices. We must, therefore, develop the necessary apparatus to deal with both the physical relationships involved in the production of biofuels, and representations of the rather opaque and intertwined policies, with the blend mandate model front row and center. To the extent that we rely on new and general models of grain and oilseed price determination and policy relationships, one might say we have taken Wright's advice with a grain of salt—or at least a "kernel of corn."

To streamline the discussion, we focus on the corn, ethanol, and gasoline markets[1] (with an assumption that the gasoline price is systematically determined by the crude oil price), noting that analogous relationships, with various nuances, hold for the markets related to sugarcane–ethanol and biodiesel.[2] We consider two core biofuel policies—blend mandates and consumption subsidies—that have been used virtually in every country promoting the consumption of biofuels.[3] We model a blend mandate because all countries use this, including the United States (see Box 2.1). Biofuel mandates and consumption subsidies by themselves do not discriminate against international trade either in the biofuel or the feedstock because they maintain access to the same market price for all.[4] The impact of trade discriminating biofuel policies such as trade barriers, production subsidies for a biofuel or a biofuel feedstock, or binary sustainability standards are discussed in later chapters.

Box 2.1 Every country implements a mandate differently

In principle, there are two types of mandates: a consumption mandate and a blend mandate. The US mandate is an annual consumption mandate (a specific volume required to be blended), but is implemented as a blend mandate. A blend mandate requires a biofuel to be a particular percentage of the fossil fuel (e.g., gasoline) or the fuel blend (e.g., gasoline plus ethanol) consumed. In fact, all countries implement blend mandates for biofuels. The only consumption mandate in the world that we are aware of is the Texas Renewable Portfolio Standard for renewable electricity.

However, each country we analyze in this book (the United States, the European Union, and Brazil) has different types of blend mandates, depending on how ethanol (E) and gasoline (G) are related [by extension, biodiesel (B) and diesel (D)].

The US blend mandate is officially announced each year by the Environmental Protection Agency at the end of November as the ratio of the *volume* of each category of biofuel (cellulosic, biomass-based diesel, advanced, and the total) and the total forecast consumption of gasoline and diesel in the following year. Because gasoline and diesel forecasts are not separated out, and because each category of biofuel includes both ethanol and biomass-based diesel, it is extremely difficult to calculate the amount of ethanol (biodiesel) mandated that is, E/G (B/D). See Chapter 10 and especially Box 10.4 for details. To analyze the economics of a blend mandate, throughout the book (for each country), we transform the ratio E/G (B/D) into E/(E + G) [B/(B + D)], which necessarily means that the transformed value of the mandate is lower than the announced value. The latter ratio is required to do a proper economic analysis of blend mandates (de Gorter and Just 2009b), it is easier to interpret as it represents the share of ethanol in the actual fuel blend consumers buy, and this ratio is required to define the "blend wall" where regular cars in the United States and Europe are unable to consume more than 10 percent ethanol in the fuel blend [25 percent for modified (non-flex) regular cars in Brazil]. It is important to understand that the "blend wall" is a blend mandate! If one is at the "wall," it means 10 percent of the fuel is ethanol. Any market shock with the blend wall requires one to analyze it as a blend mandate. See Chapter 10 for more details.

The EU blend mandate, called a "mandatory target," requires that by 2020 at least 10 percent of the total *energy* consumed in the transportation sector comes from biofuels (a tiny share of the EU mandate should come from electric cars). This is in sharp contrast to the United States where a *volumetric* share of "biomass-based diesel" and alternative sources of ethanol (cellulosic, sugarcane, and corn) are specified. This means the EU blend mandate in volumetric terms exceeds 10 percent but does not necessarily create a blend wall because the mix of ethanol and biodiesel is up to each country (or the market) to decide. Because the observed share of biodiesel consumed in the European Union is higher than the share of ethanol, the ethanol blend wall is not an issue. Although EU legislation specifies an overall blend mandate, each Member State has specified a trajectory to achieve the overall 10 percent mandate by 2020, with several Member States setting ethanol and biodiesel specific sub-mandates. However, 22 out of 27 EU Member States failed to achieve their indicative target of 5.75 percent in 2010 established under Directive 2003/30/EC (EC Commission 2013). Germany has achieved a biofuel share of 5.8 percent (European Commission 2011) on an energy basis but the European Union as a whole has achieved only 5.1 percent in volumetric terms in 2010 (USDA 2013). It should be noted that the European Union, like the United States, has carved out a minimum share of the total mandate to be second-generation biofuels. There is still debate as to how much of the total biofuel target can be first generation, derived from food crops. On July 12, 2013, the European Parliament's Environment Committee voted in favor of a 5.5 energy percent cap on first generation biofuels (UFOP 2013). There is much confusion at the time of this writing how and if the European Union will fulfil their biofuels mandate (Sandoval and Popartan 2014).

Finally, the ethanol mandate in Brazil, like the United States, is in volumetric terms, but applies to only fuel consumed by conventional cars (yet modified compared to the United States and the European Union) where the mandate varies between 18 and 25 percent (E18–E25), depending on the market situation (see Chapter 11 for details). There is no mandate for consumption by flex cars that can handle up to 100 percent ethanol (E100).

The US mandate is enforced through Renewable Identification Numbers that can be traded where a blender wishing to blend less than required can pay another firm to blend more than required (see Chapter 10 for a detailed discussion). This instrument is not used in Brazil (e.g., E25 must be 25 percent ethanol everywhere in the country) and the European Union does not use it either. As the EU's targets are to be achieved by 2020, deviations from the planned trajectory are allowed, provided they are communicated and explained to the European Commission.

This chapter begins by specifying the very specific structural relationships that determines the corn–ethanol price link. Whereas the corn–ethanol price link is immune to which biofuel policy determines the market price of ethanol, the same does not hold for the ethanol–gasoline price link. In deriving the economics of the ethanol–gasoline price link, we show ethanol prices are regime-dependent, that is, either the ethanol price is directly locked onto gasoline prices or, if the mandate is binding, the two prices become delinked. We account for the conditions that either create or break these two key price relationships.

To do so, we model the economics of blend mandate to show how ethanol prices become delinked from gasoline prices (and become negatively related) and that the causality between corn and ethanol prices changes as well. Furthermore, if abnormal profits in ethanol production exist due to ethanol production capacity constraints, the incidence on corn versus ethanol prices is also regime-dependent. If the mandate is binding, our empirical results show that abnormal profits due to capacity constraints mostly raise ethanol prices, with some reduction in corn prices (theoretically, the change in corn prices can be greater). However, if there are no mandate premiums such that ethanol prices are locked onto gasoline prices, then excess profits cause corn prices to decline with modest effects in raising ethanol prices. Excess profits in ethanol production may become the norm for the future as there are no biofuel policies to back up expansion in ethanol production capacity (see discussion in the last chapter of this book).

These two regimes means there are two states of nature in the new biofuel era: either the price of ethanol (and thus corn) is determined by the gasoline price (on a dollar per mile traveled basis), or if a mandate requires consumers use more ethanol than would otherwise prevail in the market, the mandate will drive the price of ethanol above that of gasoline. The ethanol (and corn) prices are at their lowest level possible when locked onto gasoline prices through a tax credit (an ethanol consumption subsidy); otherwise, ethanol prices float up and away from gasoline prices when the mandate determines the market price of ethanol. We provide a methodology to show when a mandate is determining the ethanol price and the magnitude of the ethanol price premium above the price that would prevail with a tax credit alone (if any).

This distinction between the two states of the nature is important because if ethanol and, therefore, corn prices are locked onto gasoline (and by extension to crude oil) prices through the tax credit, then supply/demand shocks in the corn market or changes in input costs or inventory levels have little impact on corn prices. In this case, corn market shocks could only affect prices to the extent *a change* in ethanol production or stocks affects world crude oil prices—which will be a relatively small effect indeed. The exogenous market shocks can affect corn prices only when ethanol prices float up and away from energy prices—when the mandate is binding. The extent to which these shocks influence corn prices in this case are limited by the mandate premium for ethanol prices.

This chapter is outlined as follows. The next section develops the link between ethanol and corn prices (what we call the "vertical link" in the price, quantity space). Then we investigate the relationship between ethanol and gasoline prices. This provides a basis to determine an empirical measure of how low ethanol (and thus) corn prices can theoretically go in this new era of crop–energy price linkages. This also allows us to determine the magnitude of the ethanol price premium above the tax credit due to the mandate. The following section derives the formal economic model of a blend mandate and describes its most important features. The penultimate section discusses how well the corn–ethanol price link equations perform empirically, and explains the implications for the level and volatility of corn prices. The last section provides concluding remarks.

The Link between Corn and Ethanol Prices

To derive the key link between ethanol and corn prices, we assume ethanol is produced by perfectly competitive firms using constant returns to scale technology. These are plausible assumptions given the number of plants producing corn ethanol that use a fixed proportion technology for ethanol inputs. Following de Gorter and Just (2008b) and Cui et al. (2011), we assume the ethanol processing cost per unit of output, c_0, is invariant to the quantity of ethanol produced. Empirical data for a representative plant in Iowa support this assumption as well.[5] The biggest share of processing costs is natural gas used by an ethanol plant. The assumptions here imply zero economic profits in equilibrium.[6] If the costs of producing a unit of ethanol are equal to the revenues from producing that unit (the zero profit condition), then the relationship between the price of corn, P_C, and the price of ethanol, P_E, is given by:[7]

$$P_C = \frac{\beta}{1 - r\gamma}(P_E - c_0), \tag{2.1}$$

where $\beta = 2.8$ denotes gallons of ethanol from one bushel of yellow corn, $\gamma = 0.304$ denotes bushels of dried distillers grains with solubles (DDGS) per bushel of yellow corn returned to the market as animal feed, and r represents the relative price of DDGS and yellow corn. The right-hand side term—the corn–ethanol price multiplier—before the parenthesis $\beta/(1 - r\gamma)$ has a value that has averaged approximately four and is, thus, very high. Empirically, this means that for every one cent per gallon increase in ethanol prices, corn prices increase approximately four cents per bushel.[8] Moreover, the higher the value of the co-product, the higher the benefits of ethanol production and, as a result, producers are willing to pay more for corn. This would lead ethanol prices to have a greater impact on corn prices. The average value in the period 2007–2014 of the multiplier is 3.91 (with a monthly high of 4.82 and a low of 3.56), and the average value of r is 0.93 (with a high of 1.38 and a low of 0.70). The highs of each of these two parameters occurred in 2013 and 2014, which has implications for the model's prediction error, a discussion of which is delayed to a later section in this chapter.

The formula linking corn and ethanol prices should be augmented to account for the extraction of corn oil from DDGS that has become increasingly common. The corn oil is subsequently sold for biodiesel production. Denoting θ as pounds of oil from one bushel of corn and P_{CO} as the price of corn oil, the new corn–ethanol price linkage equation becomes:

$$P_C = \frac{\beta}{1 - r\gamma}(P_E - c_0) + \frac{\theta}{1 - r\gamma}P_{CO}. \tag{2.1'}$$

Comparing average predicted prices of corn from equation (2.1) versus (2.1′), allowing for oil extraction from DDGS adds 18 cents per bushel on average to the predicted price. Notice that the equation (2.1′) implies that willingness to pay for corn also depends on the biodiesel price. This is true especially when the price of biodiesel is very high, elevating the corn oil price, which in turn may exert substantial influence on the price of corn. Paradoxically, a commodity price can be determined by a seemingly unrelated product—the biodiesel price. The combination of these factors suggests that we cannot pin down the exact relationship in equations (2.1) and (2.1′) empirically, although the presented approximations are useful and revealing. It is also interesting to note that removing the corn oil from DDGS can lower the relative price of DDGS (in other words, the parameter *r*), although hog and dairy farmers prefer less oil in the DDGS for the feed ration to prevent excess fat in pork bellies and milk suppression, respectively.[9] But ultimately, one has to look at the entire corn–ethanol and soybean oil–biodiesel complex as DDGS competes more with soybean meal than corn (DDGS has more protein like soybean meal and corn has more energy).

Moreover, corn and soybeans are like joint products in production, with crop rotations typically involving both crops; and, soybean meal and corn can also be viewed as complementary in feed rations. Thus, there are multiple avenues in how ethanol and biodiesel policies increase the level and volatility of food commodity prices. As we will see later, EU and Brazilian policies also have implications for affecting commodity prices (and these interact with US policies).

We would like to note that neither equations (2.1) nor (2.1′) implies a strict causal relationship between ethanol and corn prices. The direction of causality depends on the policy determining the ethanol market price: if the tax credit is binding, then the ethanol price (in conjunction with the gasoline and hence crude oil price) causes the corn price.[10] However, we will show later in developing the blend mandate model that ethanol and corn prices are determined simultaneously and, therefore, the causality is ambiguous with a binding mandate.

The Link Between Ethanol and Gasoline Prices

So far we have determined the corn–ethanol price link. The second task before us now is to understand the relationship between ethanol and gasoline prices. This relationship will allow us to determine which policy regime prevails. If the tax credit is binding, ethanol prices will be at the lowest possible level they can go (barring capacity constraints and other market failures) whether supply or demand shocks are present or not. However, if ethanol prices float up and away from the lowest levels they can go, then we know the mandate is binding. We can also use the gasoline–ethanol price relationship to determine the mandate premium above the tax credit.

Using standard economic theory, we assume consumers of fuel (a blend of ethanol and gasoline) obtain utility from the number of miles traveled. Denote the *volumetric* fuel tax (charged on a gallon of fuel) by t. In this case, the fuel tax per mile for each fuel is t divided by the miles per gallon of the fuel. Thus, because the number of miles per gallon of ethanol is only approximately 0.70 times the miles per gallon of gasoline (see also Box 2.2), the tax per mile traveled using ethanol is about 1.42 times higher than that for gasoline.[11] Define λ as the ratio of miles per gallon of ethanol to miles per gallon of gasoline, where $\lambda = 0.70$.

Box 2.2 How many miles can you get from a gallon of biofuel?

There is much confusion as to how many miles per gallon one can obtain from a gallon of ethanol (biodiesel) relative to a gallon of gasoline (diesel). Most economists use the relative energy content of the biofuel to its mother fuel (0.666 and 0.931 for ethanol and biodiesel, respectively). Thus, the term "gasoline energy-equivalent gallon" (GEEG) is often used in the literature. However, there is not a one-to-one correspondence between the energy content and miles traveled. For example, gasoline has an energy content that is only 0.905 times that of diesel yet diesel cars on average get approximately 34 percent more miles traveled per unit volume than gasoline (Belzowski and Green 2013).

Ethanol gets approximately 70 percent of the miles of a gallon of gasoline on average, but, as for diesel cars, it varies by the car type and various characteristics like altitude, temperature, and even how much air is in your tires. Likewise, biodiesel is found to get 0.913 miles per gallon compared to a gallon of diesel. Therefore, we prefer to use the term "miles equivalent gallons of gasoline" (MEGG) or "kilometers equivalent liters of gasoline" (KELG). We also need to calculate the miles per gallon of fuel used in US "flex cars" that are able to consume any blend between zero (E0) and 85 percent ethanol (E85). This is critical information to determine the parity price for E85 relative to E10, which is the maximum amount of ethanol regular cars can consume. Applying the methodology of Babcock and Pouliot (2013), but using 0.70 for the relative mileage of ethanol and gasoline instead of two-thirds as in the original study, the E85 price at parity is 0.768 times the price of E10 (as opposed to 0.741 in Babcock and Pouliot 2013).

It is important to understand how many miles one gets per unit volume when calculating how low ethanol prices can go, as shown in equation 2.2, and analogous equations derived for the biodiesel–diesel price relationships in Chapter 9 and for the price links for two types of ethanol used in Brazil (Chapter 11). Furthermore, this information is important when we assess how efficient biodiesel is relative to corn-ethanol in Chapter 9. Curiously, the European Union has an overall biofuel mandate (no specific volumes for ethanol versus biodiesel that appears to be the case in the United States, although Chapter 10 on the complex structure of the US mandate will show some nuances) that is expressed in "energy equivalent" terms but biodiesel gets a weight 57 percent higher than ethanol although biodiesel gets 75 percent higher miles per gallon than ethanol.

The maximum price that consumers are willing to pay for ethanol is λ times the consumer price of gasoline, $\lambda(P_G + t)$, where P_G is the market price of gasoline.[12] If the ethanol price is above this price, consumers could purchase the same distance traveled by using gasoline exclusively. In the absence of other policies, market forces will render the price of a mile using ethanol and the price of a mile using gasoline to be equal so long as both fuels would be produced. This relationship implies that the wholesale price of ethanol is given by:

$$P_E = \lambda P_G - (1 - \lambda)t. \tag{2.2}$$

Refiners and blenders are required to pay the fuel tax on both ethanol and gasoline by volume. Nevertheless, consumers value ethanol for its contribution to miles and, therefore, are willing to bear a smaller portion of the fuel tax for ethanol than for gasoline. Thus, the market price blenders are willing to pay for ethanol is lower by $(1 - \lambda)t$, the "penalty" on ethanol due to the fuel tax being levied on a volume basis rather than on a miles traveled basis. This tax thus leads the market to favor gasoline more than would be suggested by the simple advantage in miles per gallon.[13] The implicit penalty on biofuel production can be very significant in countries like the United Kingdom, which has a fuel tax of $3.40 per gallon, thus imposing a penalty on ethanol producers of $1.02 per gallon.[14] Therefore, the free market ethanol price is very low. In general, one expects the correlation between ethanol and gasoline prices to be highly positive when the tax credit is binding.

We now introduce a volumetric ethanol consumption subsidy such as the US blender's tax credit t_c. To take advantage of the government subsidy offered them, blenders of ethanol and gasoline will bid up the price of ethanol (moving up the supply curve for ethanol) until it is above the predicted wholesale price of ethanol given in equation (2.2) by the amount of the tax credit. Otherwise, blenders would be foregoing money represented in the tax credit. The wholesale price of ethanol is now given by:

$$P_E = \lambda P_G - (1-\lambda)t + t_c. \tag{2.3}$$

Because fuel consumers are indifferent to using gasoline and using a mileage equivalent amount of ethanol, and because gasoline prices change little with changes in ethanol production, the incidence of the tax credit is such that ethanol producers derive most of the benefit.[15] However, when the mandate is binding, the wholesale price of ethanol will be determined by the point on the ethanol supply curve that corresponds to the required level of production to fulfill the blend mandate. In this case, equation (2.3) will not hold, but a higher price of ethanol will be realized (see Box 2.3).

Box 2.3 What exactly is a "mandate price premium"?

A "mandate price premium" is simply the difference between the observed ethanol price and the hypothetical (calculated) price under the tax credit (if any) described in equation (2.3). In our discussion of Figure 3.2 in Chapter 3, we will show that this premium was negative in four brief time periods in the past because of ethanol-blending constraints. However, if positive, then consumers are paying more than they wish, therefore, ethanol consumption is "mandated" upon them [we reject the theory of Anderson (2012) that consumers desire to pay more for ethanol than the mileage equivalent value *vis-à-vis* gasoline]. Paying more for ethanol on a miles per gallon basis represents "no choice." This price premium above that generated by the tax credit alone can be a result of several situations, including the formal mandate itself; de facto mandates in the form of environmental regulations; when consumers are unable to choose between ethanol and gasoline at the gasoline pump on a miles obtained basis because of a lack of flex cars and E-85 stations, so ethanol may be sold at a premium; or when prices are determined in world markets where biofuel policies in the rest of the world are determining the ethanol price premium. See also Box 1.2 "Do economists agree on what a biofuel policy is?"

Before analyzing the historical data to determine when the tax credit was binding and the magnitude of ethanol premiums over the tax credit when the mandate was binding, we first introduce an economic model of blend mandates.

An Economic Model of a Blend Mandate

Consider a blend mandate $\alpha = C_E/(C_E - C_G)$, where C_E is the consumption of ethanol and C_G the consumption of gasoline, and α is the required minimum percentage of ethanol in fuel consumption (the ethanol–gasoline mixture). We present a simplified model of a blend mandate in Figure 2.1 where an exogenous crude oil-based gasoline price is assumed, a single supply of gasoline curve (no distinction between domestic and imported gasoline supply), and no imports of biofuels.[16]

Consider a competitive market with a domestic supply curve for ethanol, S_E, and a supply curve for gasoline, S_G, as in Figure 2.1. The domestic demand for fuel is denoted by D_F. Ethanol and gasoline are assumed to be perfect substitutes in consumption. For ease of exposition, the intercept of the ethanol supply curve S_E is arbitrarily set to coincide with the price of gasoline on a miles-equivalent basis.

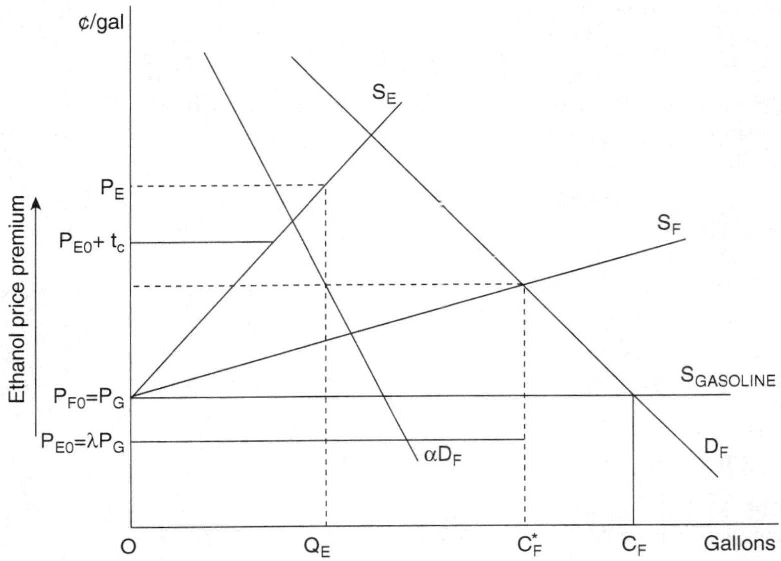

Figure 2.1 Economics of a biofuel blend mandate.

The mandate drives up the price of ethanol relative to an equilibrium under no mandate, causing the market prices for ethanol and gasoline to diverge. In equilibrium, consumers must pay the marginal cost to the blender of producing a unit of fuel mixture. This marginal cost is given by the weighted average price of ethanol and gasoline where the weights are formed by the required share of ethanol under the mandate:

$$P_F = \alpha P_E + (1 - \alpha) P_G, \tag{2.4}$$

where P_F is the consumer-weighted average price, P_E is the corresponding market supply price of ethanol, and P_G is the price of gasoline. This constitutes the marginal cost of the mandated mixture because P_E is the marginal cost of ethanol and P_G is the marginal cost of gasoline to the blender. To find the market equilibrium prices P_E and P_F, we must determine the market prices that cause total fuel supply to equal total fuel demand. This requires the derivation of a total fuel supply curve, $S_F(P_F)$, determined by the component supply curves S_E and S_G. Because our assumption of an exogenously determined gasoline price and equation (2.4) implies a one-to-one relationship between P_E and P_F, we can represent the ethanol supply curve S_E as a function of P_F.

With a biofuel mandate requiring α percent of the total fuel to come from ethanol, the demand for ethanol at any fuel price P_F is given by the curve αD_F. The market price of fuel P_F is determined where the marginal cost curve of fuel blenders, S_F, intersects the fuel demand curve D_F. When the price and quantity of fuel are determined, the quantity of ethanol, Q_E, can readily be found. Finally, given the ethanol quantity, the ethanol price, P_E, is read from the corresponding point on the ethanol supply curve.

Because no taxpayer monies are involved with mandates, the consumers fund the additional production of ethanol with the price they pay at the pump, resulting in higher fuel prices and higher ethanol prices. A blend mandate is, thus, equivalent to a combination of a fuel tax and a biofuel production subsidy. The extra revenues garnered by biofuel producers due to the mandate are given by $P_E \times Q_E$, but consumers only value it by $P_{E0} \times \lambda P_G$. The transfer from consumers is $(P_E - P_{E0}) \times Q_F$.

The Importance of the Blend Mandate Model

Not only does every country, including the United States, implement a blend mandate, the "blend wall" is also a (maximum) blend mandate. We are now in a position to summarize some key features of the blend mandate model that will prove useful in the analysis to follow in this book:

(1) If there is an exogenous increase in the gasoline price, then fuel consumption declines along with ethanol prices because less ethanol is consumed and, therefore, produced. This means that the ethanol and gasoline prices are negatively correlated under a blend mandate.
(2) A corn supply/demand shock will have a large impact on corn–ethanol prices because the demand for ethanol (with respect to ethanol prices) is very inelastic, the production of ethanol is essentially fixed by the mandate to the quantity of gasoline, and the latter does not respond much to the shock in the corn market.
(3) The effect of a change in the blend mandate on fuel prices is ambiguous (unlike the simple case depicted in Figure 2.1 of a fixed gasoline price where consumers necessarily pay higher fuel prices) because, with an upward sloping supply curve for gasoline, gasoline, and, thus, crude oil prices decline with ethanol supply. The net effect of a mandate on fuel prices depends primarily on relative supply elasticities of ethanol versus gasoline (de Gorter and Just 2009b). Therefore, the mandate's impact on corn prices is thus also ambiguous (Drabik 2011). The outcome depends on how much gasoline consumption declines with the ethanol mandate; if by a small amount, then ethanol and corn prices increase but, alternatively, if the increased mandate for ethanol increases prices to the point that fuel consumption declines severely, the increased mandate may result in less consumption of ethanol. In this case, the increase in the mandate may result in a reduction in corn prices. This latter case is probably less common but still plausible if the costs for additional biofuel production is steeply increasing. Thus, there are conditions where ethanol and corn prices decline with changes in the mandate levels (Drabik 2011, 49).
(4) An exogenous change in the blend mandate requirement has an ambiguous effect on the ethanol price and, thus, also on corn prices when the gasoline price is determined by market forces (i.e., is endogenous) (Drabik 2011). Intuitively, because a change in the blend mandate has an ambiguous effect on the fuel price, it must also affect the fuel quantity ambiguously. However, fuel and ethanol quantities are firmly linked through the mandate which means that ethanol quantity can increase or decrease with a higher mandate, implying that the effect on the ethanol price is also ambiguous. The effect on the corn price follows directly from equation (2.1).
(5) Because ethanol prices are simultaneously determined with gasoline prices (when the supply of gasoline/crude oil is upward sloping, unlike that depicted in the simple case in Figure 2.1), there is no strict causality between corn prices causing ethanol prices or vice versa, but we know that the ethanol demand curve is very inelastic

(with respect to the ethanol price), therefore, corn prices cause ethanol prices much more than vice versa, especially with low values of the blend mandate parameter α. To see this, consider a quantity mandate which stipulates a fixed amount of ethanol to be consumed (in other words, the ethanol demand curve is perfectly inelastic). Thus, a shock in the corn market will affect the total quantity of corn (with the amount of corn devoted for ethanol unchanged) and, therefore, its price. The change in the corn price, in this particular case, then translates into the ethanol price [as per equation (2.1)], illustrating our point that under the mandate, the corn prices cause ethanol prices much more than the other way round.

(6) The incidence of excess profits in ethanol production due to ethanol production capacity constraints (a fixed wedge, for a given constraint, between corn and ethanol prices) with a binding mandate will have ethanol prices going up far more than corn prices going down. For a given blend mandate, the size of the ethanol market is determined by the size of the gasoline market. With the excess profits, competition among ethanol producers will prevent the corn price to be bid up because ethanol quantity is essentially fixed by the size of the gasoline market. Therefore, it can be expected that the ethanol producer will rather keep lowering the ethanol price to capture a higher market share until the excess profits dissipate. However, if there are no mandate premiums and, thus, ethanol prices are locked onto gasoline prices, most of the adjustment in eliminating excess profits through ethanol capacity expansion would be in higher corn prices.

Nevertheless, the conclusions in (6) here assume there is no regime change in each case. For example, if the mandate is binding before profits erode, it may not be binding after zero profits are achieved. When analyzing regime change with changes in profits, there are two possible starting points:

(1) Zero profits. In that case, either the mandate or the tax credit can be binding. If there are capacity constraints that drive a wedge between corn and ethanol prices, then ethanol prices never decrease. But, a higher ethanol price with a previously binding tax credit can hardly be attributable to a mandate. Thus, there is no regime reversal in this case nor when profits are eliminated because, then, we are back to the situation with the initially binding policy.

(2) Positive profits. If the tax credit is binding to start with, then when the profits are eliminated, the ethanol price decreases and, thus, there is no regime reversal, because the ethanol price will be even lower than before. However, if the mandate is binding with positive profits, then the elimination of the profits can lead to a regime reversal.

How Accurate is the Theory? Some Numerical Results

We have derived equations that linked corn to ethanol prices and ethanol to gasoline prices, along with a general theory of blend mandates with important market impacts that differ from a binding tax credit. Three natural questions arise: how well does the corn–ethanol price link theory predict actual prices [equation (2.1)]?; how low could corn prices have gone under "free market" ethanol prices when locked onto crude oil prices [equation (2.2)]?; and what were the ethanol price premiums due to the mandate above and beyond the tax credit [in other words, what were the observed ethanol prices minus those predicted by equation (2.3)]? The empirical results for the first two questions are presented below in Figures 2.2 and 2.3, respectively. The question of ethanol price premiums due to tax credits versus mandates requires more theory before we can adequately address the issue empirically and, thus, empirical estimates are delayed until the next chapter.

How Well Does Our Theory of the Corn–Ethanol Price Link Hold Up Empirically?

One would normally assume that the assumptions underlying equations (2.1) and (2.1′) are justifiable in the long run when the industry is likely to be in equilibrium, but it is less probable to hold in the short run. In other words, it can be expected to have corn prices predicted by our theory to deviate from the observed values but only in the short run. Figure 2.2 displays the predicted corn price from equation (2.1) (using data for a typical ethanol plant in Iowa), the actual corn price, and prediction error (the difference between predicted and actual corn prices).[17]

As expected, there are prediction errors over time that do not typically last long; the one very important exception is the recent time period beginning in August 2013 that has persistently exhibited abnormal profits but have collapsed in September 2014. To quote Irwin and Good (2014): "The extended period of ethanol production profitability appears to defy basic economic logic…Economic logic suggests this situation is unsustainable." We delay our discussion of the recent time period and first explain the reasons for prediction errors in the earlier time periods.

As summarized in the lower graph in Figure 2.2, the significant errors identified are mostly over predictions (peaking at 27, 34, 36, and 96 percent error in January 2008, November 2009, November 2011, and May 2014, respectively). The over predictions are due to ethanol production-capacity constraints. Over predictions imply excess profits in ethanol production and under prediction implies negative profits, both undermining the zero profit condition that underpins equation (2.1). The first major prediction error

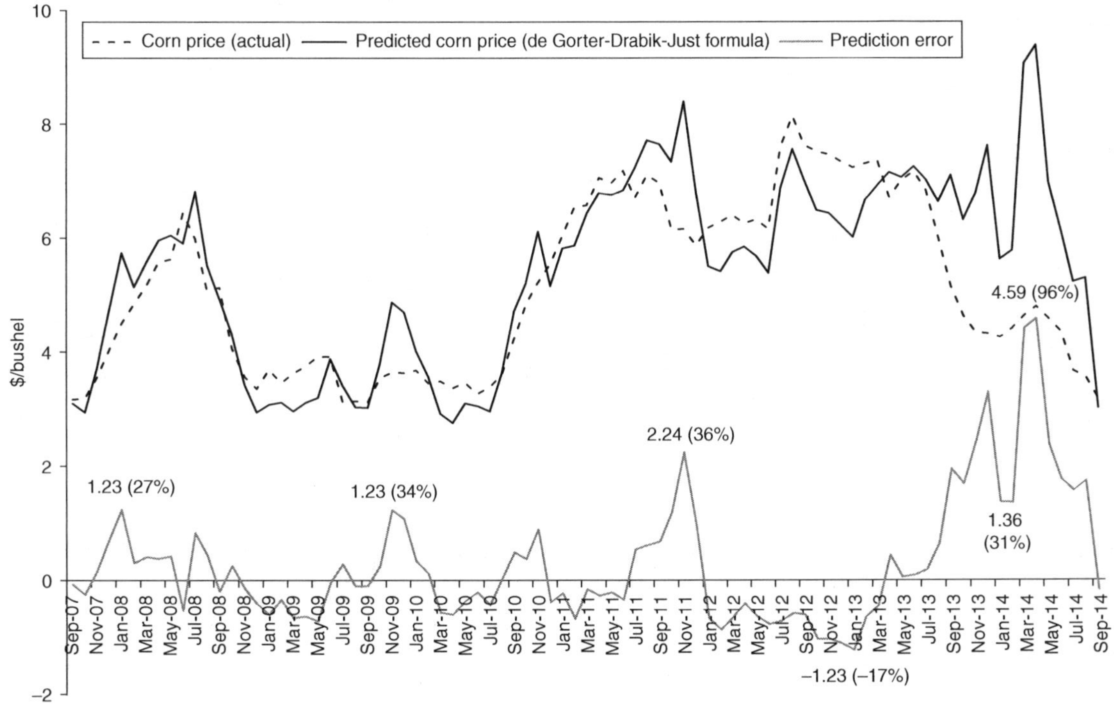

Figure 2.2 Predicted versus actual corn prices.

Source: Calculated using data from CARD (2014a), AMRC (2014), and EIA (2014a).

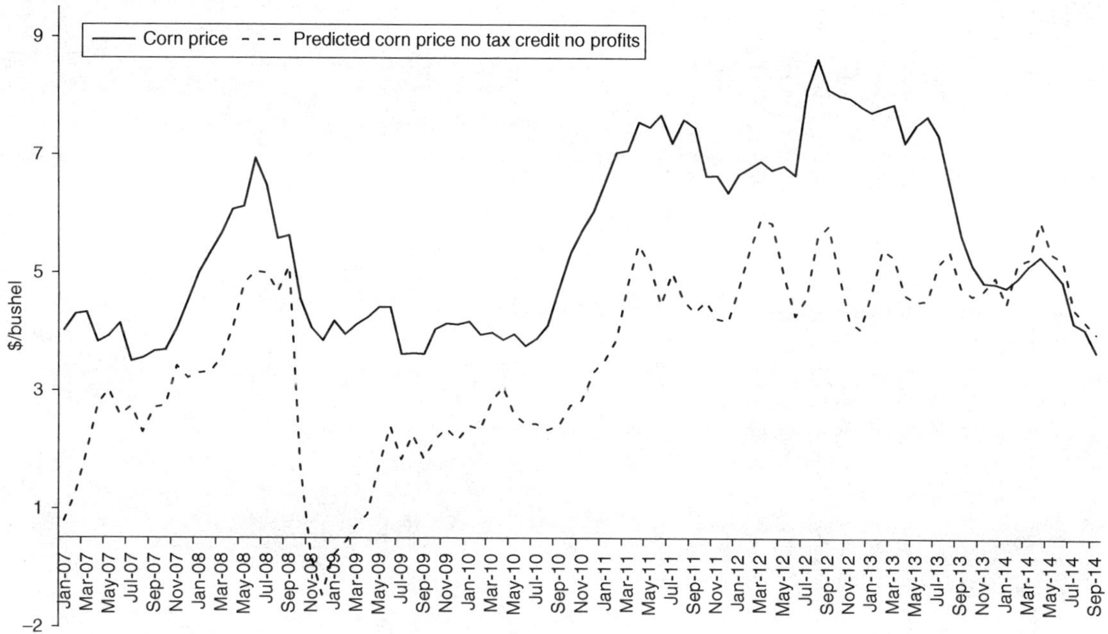

Figure 2.3 How low can corn prices go? (historical)

Source: Calculated using data from CARD (2014a) and EIA (2014a).

of January 2008 reflects the effects of a ban on MTBE as well as a spate of high oil prices. These, together with the tax credit creating the link between ethanol prices and corn prices, resulted in capacity constraints in ethanol production, allowing for excess profits in the short run until investments in new capacity came to fruition. The prediction error in the second instance (November 2009) was likely due to fall out from the financial crisis and low gasoline prices that caused huge declines in share values of ethanol companies and an unusual number of ethanol plant bankruptcies in 2008–2009. These plants were eventually bought up (for dimes on the dollar), but the financial restructuring took time before production could get back online.

The period of severe over prediction corresponding to November 2011 when export demand was high was created by a pair of poorly crafted policies. A loophole in the import tariff imposed by the European Union was allowing US ethanol exports to obtain a higher price. Meanwhile, the domestic tax credit in the United States helped subsidize ethanol production in the form of high domestic ethanol prices with exports. Ethanol production-capacity constraints ensued. The European Union closed the import tariff loophole in November 2011 and the US tax credit expired at the end of 2011, eliminating any capacity constraints.[18]

The largest under prediction peaked at 17 percent in January 2013. Production had to be curtailed from the fallout of the tax credit expiring and the EU import tariff loophole closed. Negative profits ensued and, thus, the under-predictions in the period marked by January 2013. These under-predictions (and negative profits in ethanol production) continued through 2012 due to the US corn drought.

However, since August 2013, profits in ethanol production have soared and so too the over predictions of equation (2.1). There have been multiple reasons for the excess profits in this extended time period.[19] The relative price of DDGS reached record levels (excess demand for protein and high soybean meal prices; no import tariff for DDGS into China, unlike for corn) whereas the ethanol price rose due to transportation bottlenecks (caused by a bad winter and a shortage of rail cars due to expanded supplies of shale oil); in other words "marketing" constraints. Ethanol prices rose and corn prices declined in face of a record corn crop in 2013–2014. Meanwhile, export volumes and prices for ethanol remained robust.

These excess profits have disappeared for the moment. But, will they return and, to quote Irwin and Good (2014), "defy basic economic logic" and are "unsustainable"? We are unsure. Ethanol production in 2014 has been running at record levels and at maximum capacity. Even with record profits since August 2013, no new ethanol plants are under construction. There is only one ethanol plant being built.[20] It appears that the industry is reticent about expanding ethanol production capacity. Several reasons can explain this. Corn–ethanol production capacity is close to 15 billion gallons, equal to the residual mandate or "renewable fuel gap" of 15 billion gallons in 2015 for which corn-ethanol is eligible (Chapter 10 explains that there is no "corn-ethanol mandate" per se). Thus, any expansion in ethanol capacity has no mandate backing it up. Indeed, as we show in Chapter 10, the implicit mandate for corn–ethanol is only approximately 13 billion gallons; not 15 billion gallons. Corn–ethanol exports account for the difference. In fact, the US Environmental Protection agency (EPA) announced in November 2013 that the biofuel mandate was to be scaled back 1.4 billion gallons because of the ethanol blend wall, and this has not yet been resolved at this time of writing (September 2014). The industry faces much policy uncertainty.

Unlike in the past, there is no tax credit. Thus, there is no biofuel policy that undergirds expanded corn–ethanol production (unless one counts foreign policies that account for US ethanol exports). The situation in 2013–2014 exhibits the power biofuel policies had in the past, allowing for continually expanding capacity and to allow for zero profits and equation (2.1) to hold except for brief periods of time. However, the situation going forward is very different than in the past, reflected in persistent excess profits in recent times. Will persistent excess profits return? We delay our assessment of the future for biofuels and food commodity prices until the concluding chapter of this book.

In addition to ethanol production-capacity constraints, excess profits from September 2013 to May 2014 are also due to record highs in the relative price of DDGS and corn—denoted by r in the equations earlier. A short-term increase in the relative price of DDGS makes the ethanol co-product more valuable at prevailing corn and ethanol market prices, thus giving rise to excess profits. The question then becomes what the market incidence of a high r will be. It can be expected that high values of r will stimulate ethanol production, requiring more corn to be processed. The price of corn will, therefore, increase, reducing the profits and establishing the zero profit condition for ethanol production.

There are, therefore, three sets of historical situations our book sets out to explain in terms of prediction errors.

(1) Zero profits:
- Tax credit binding (especially in 2008)
- Mandate binding

(2) Positive profits
- Short periods of time in the past, except since August 2013, when excess profits have become seemingly perpetual

(3) Negative profits (2012 to early 2013) when our model under predicts corn prices

It should be noted that when we write that our model over predicts the price of corn with positive profits, it has to be interpreted with care. If the tax credit is binding, then elimination of the excess profits (capacity constraint) will have actual corn price rise to predicted values (so our predicted price is close to what it should be with no profits). But that is not the case if the mandate is binding—our predicted price is a true over prediction if there were no profits.

How Low Can Corn Prices Go?

Using equation (2.2) where we assume no ethanol policies and consumers' value miles traveled and face no ethanol blending constraints, we can calculate the lowest prices ethanol can go. Again, we assume zero profits in converting ethanol prices into corn prices using equation (2.1). These predicted prices are plotted against observed corn prices in Figure 2.3.

The first thing to notice is the negative values in late 2008 and early 2009. This is because gasoline prices were so low in that time period that corn prices are required to go negative to maintain the zero profit condition. Clearly, this is not realistic and, therefore, one has to adjust the theory (and, thus, the data) to comply with the reality that there exists a "shut-down" ethanol price at positive corn prices (the intercept of the ethanol supply curve is above the demand for ethanol).[21] We make these adjustments in Chapter 3 in order to determine the effect of biofuel policies on food commodity prices. In general, Figure 2.3 shows that actual historical corn prices are very high compared to the lowest ethanol (corn) prices could have gone (assuming no ethanol policies).

The second thing to notice in Figure 2.3 is that, since November 2013, actual corn prices are about equal to the free market or lowest corn prices can go [as predicted by equation (2.2) in combination with equation (2.1) assuming zero profits in ethanol production]. Does this mean current corn prices are at free market levels for the first time ever? The answer would be yes if actual profits in ethanol production were zero, but we know this has not been the case except for the month of September 2014. To understand this better, consider Figure 2.4 that shows the predicted corn prices with and without profits for the recent time period.

With profits, the predicted corn price [using equation (2.2) with equation (2.1)] is much lower and is even negative in March 2014, whereas actual and predicted corn prices with zero profits track each other quite closely. The reason for the latter is that the mandate premium over the free market ethanol price offsets the profits in ethanol production. Furthermore, the ethanol price is predicted from gasoline prices in equation (2.2) and

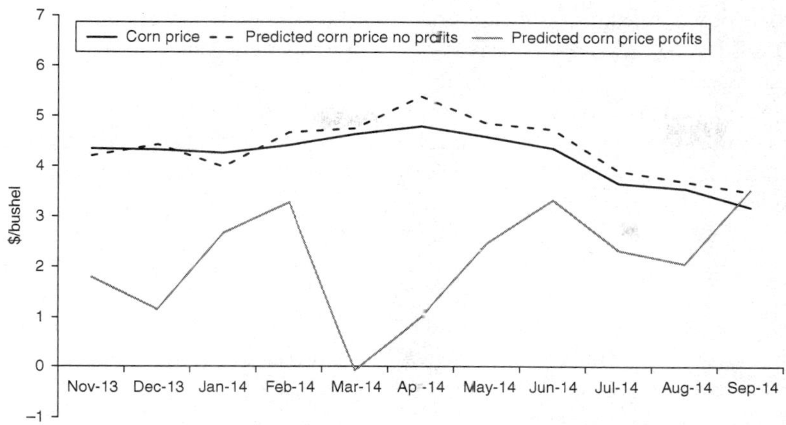

Figure 2.4 How low can corn prices go? (recent months)

Source: Calculated using data from CARD (2014a), AMRC (2014), and EIA (2014a).

so diverges from actual ethanol prices. Thus, a combination of mandate premiums and divergence between actual and predicted ethanol prices makes actual corn prices appear as if they coincide to free market corn prices in Figure 2.4 when they really do not except for the last value. For September 2014, the predicted corn price in Figure 2.4 is $3.53 per bushel, yet the actual corn price is $3.18 per bushel. With zero profits, this is unexpected, but actual ethanol prices are below that dictated by equation (2.2). Ethanol is being sold at a discount to gasoline on a miles per gallon basis, probably because of the blend wall. We discuss this in further detail in Chapter 10.

Concluding Remarks

The key contributions of this chapter are (1) to derive the link between corn and ethanol prices; and (2) to derive the link between ethanol and gasoline prices. In the process of developing the concepts of corn–ethanol and ethanol–gasoline price links, we have been able to disentangle the impact of the corn ethanol tax credit from the mandate in a closed economy setting. The tax credit is part of the link between ethanol and gasoline prices, resulting in the first of two states of nature we summarized in Chapter 1: ethanol (and thus corn) prices are *lowest* when *locked* onto gasoline prices plus the tax credit. Otherwise, mandates impact the ethanol price by delinking it from the gasoline price, with mandate premiums, generating the second state of nature, namely, ethanol prices *float up and away* from gasoline prices. Therefore, we analyze the economics of the tax credit and the blend mandate separately. The analysis of the all-important interaction effects not only between ethanol tax credits and mandate but also between other biofuel policies, especially (feedstock and biofuel) production subsidies, is delayed until Chapter 12. Now, we are also able to determine the free market price of ethanol.[22]

The production process for ethanol has great importance in explaining the recent rise in food commodity prices. To explain the key price and quantity linkages in the feedstock, biofuel and fuel markets, we first develop the corn–ethanol price link. In modeling this link, it is important to take into account both the role of an ethanol *co-product*, DDGS.[23] DDGS is typically used to feed livestock and is, thus, a close substitute for corn. Moreover, corn oil can be extracted from DDGS and used to produce biodiesel. The proportional amounts of each substance created in the production process, as well as the standard assumptions regarding a competitive marketplace for outputs and inputs, results in a very specific corn–ethanol price relationship. We show this tight price link, represented by the corn–ethanol price multiplier, is to hold in theory with the zero profit condition regardless of the biofuel policy determining the biofuel prices.[24] We provide empirical evidence of the predictive power of this theoretical price equation and the reasons for prediction errors.

An important concept that we later use to explain the impacts of biofuel policies on the rise in the food commodity prices is the free-market ethanol price. This is a hypothetical price that would prevail in the market in the absence of any biofuel policy and that would make consumers indifferent between buying ethanol and gasoline. This price takes into account the differential mileage the two fuels produce as well as the existing fuel tax. Thus, at this hypothetical price, traveling a mile using ethanol costs the same amount as traveling a mile using gasoline. Because a gallon of ethanol generates 30 percent fewer miles traveled than a gallon of gasoline, we show that the volumetric fuel tax (charged on a per gallon rather than a per mile basis) penalizes ethanol blenders. This is one reason why the free market ethanol price has historically been below the shutdown price on the ethanol supply curve (the price below which it is no longer profitable to produce ethanol). Ethanol production would not have occurred without biofuel policies. This implies that a part of the increase in the ethanol price, due to biofuel policies, above its free-market level was eaten up by the gap between the shutdown price (given by the vertical intercept of the ethanol supply curve) and the free-market ethanol price. We term this gap the "water" in the ethanol price premium and discuss it in further detail in the next chapter.

The literature that continually assumes nothing has changed and proceeds with the use of the old tired models to explain food commodity price formation in this new era loses relevance in that they do not have the power to properly measure the impact of biofuels policies, capacity constraints, or crude oil market fluctuations on oilseed and grain markets. Thus, the standard analysis of sectoral supply/demand shocks and depleted stocks causing the unprecedented price booms is misplaced; in essence, this squarely ignores the realities of this new biofuel era.[25] We, therefore, now proceed to outline this game-changing theory of food commodity price determination that outlines the core concepts required in understanding all other topics described in the following chapters, including how biofuels policies have affected the level and volatility of grain and oilseed prices since 2006, the interaction between policies and countries, and a critique of the alternative literatures.

CHAPTER 3

MEASURES OF BIOFUEL POLICY IMPACT ON FOOD COMMODITY PRICES

The previous chapter developed the required analytical framework in order to determine how much of the food grain and oilseed price increase observed over the last several years is due to biofuel policies. We established the price links between corn and ethanol, and between ethanol and gasoline. However, we failed to provide empirical estimates of the ethanol price premiums due to the tax credit versus the mandate; these premiums were far too high. Our predicted monthly corn prices went far too low (and even negative). Using annual data, Drabik (2011) estimates that actual ethanol price was, on average, 3.6 times that of the estimated free market levels as per equation (2.2). Therefore, historical ethanol policy price premiums have to be adjusted to account for the ethanol shutdown price (at positive corn prices). In other words, we have to first determine the intercept of the ethanol supply curve and then adjust the data for the redundancy in ethanol price premiums (which we call "water" in a biofuel price premium). This is the first order of business in this chapter; namely, derive the ethanol supply curve and its intercept by analyzing the link between corn and ethanol quantities (the "horizontal link" to complement the "vertical" price link derived in the previous chapter), including the "recycling effect" of DDGS.

Once this analysis is completed, we then undertake several empirical exercises to estimate:

(1) The level of ethanol price premiums due to the tax credit versus the mandate adjusted for "water"
(2) The change in corn prices due to ethanol policies under two different counterfactuals (a) the reality of the new corn–ethanol price link; and (b) the good old days when biofuel policies and the corn–ethanol price link did not exist

(3) The transfers from grain/oilseed consumers to producers
(4) The cost savings to consumers and taxpayers of allowing the US tax credit to expire, given that corn prices may in the near future become locked onto crude oil prices (through ethanol and gasoline prices) with no excess profits in ethanol production and no mandate premiums for ethanol prices

Our empirical analysis in this chapter finds that (1) ethanol price premiums due to biofuel policies are high even after adjusting for redundancies; (2) 80 percent of the corn price increase in the period 2007–2011 would have occurred anyway, given the counterfactual of the corn–ethanol price link with biofuel policies putting a variable price floor under corn prices; (3) 40 percent of the corn price is due to ethanol policy compared to the counterfactual of no biofuels (unadjusted for nonlinear price effects due to low stock levels due to biofuel demand shock); (4) biofuel policies represent yet another $1 billion per day farm subsidy [in addition to that estimated by the Organization of Economic Co-operation and Development's (OECD's) "producer support estimate" (PSE)]; and (5) allowing the tax credit to expire may save over a billion dollars per day transfers from grain/oilseed consumers to producers, should corn prices become lowest when locked onto crude oil prices in the future.

Nevertheless, before we undertake the empirical analysis, we have to deduct the element of redundancy in the ethanol price premium due to biofuel policies; for this, we need to determine the ethanol supply curve and its intercept.

Deriving the Ethanol Supply Curve and Redundancy in Ethanol Price Premiums

Consider the corn market depicted in the first panel of Figure 3.1. The supply of corn is given by the curve S_C. The demand for corn outside of ethanol is given by D_{NE}, where NE represents "non-ethanol" uses. The latter represents aggregate (domestic and export) demand for feed/food corn facing US corn producers. The intersection of these two curves gives the "non-ethanol" corn price P_{NE}, which represents the intercept of the ethanol supply curve in bushels of yellow corn.

Define P_G and P_{E0} as the free-market prices for gasoline and ethanol, respectively [where P_{E0} is derived from equation (2.2) in Chapter 2]. The actual price of ethanol P_E is higher than P_{E0} by the ethanol price premium due to either a mandate or tax credit. Corresponding to the ethanol price P_E is the corn price P_C, equal to the price of ethanol in dollars per bushel, P_{Eb}. The ethanol price P_E can be converted into P_{Eb} using equations (2.1)

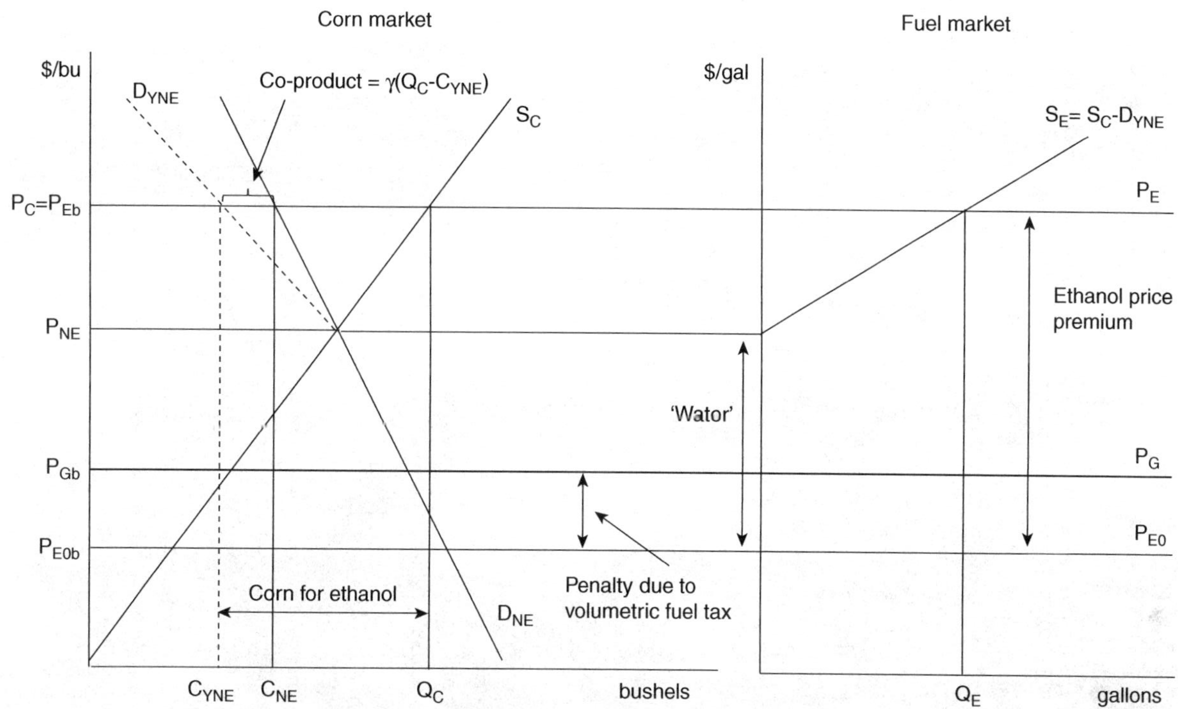

Figure 3.1 The ethanol supply curve and "water" in the ethanol price premium.

Source: Drabik (2011).

or (2.1′).[1] If the corn market price is linked to the ethanol price through equation (2.1) and the ethanol price is linked to the gasoline price (which is linked to the crude oil price) through equation (2.3)—as is the case when the US tax credit is determining the ethanol price—then any supply/demand shifts have little effect on the corn price.[2] The only impact is due to the resulting change in gasoline and crude oil prices, which are likely very small. The only effect of supply or demand shifts when ethanol prices are tied directly to crude oil prices through the tax credit is to change the (unobserved) non-ethanol corn price, P_{NE}.

Nevertheless, a change in P_{NE} results in a change in the level of ethanol production, even without biofuel policies should "water" not exist. This point seems to be forgotten in the debate about the role of ethanol policies on corn prices as most economists only calculate the "breakeven" crude oil price as if that represents the intercept of the ethanol supply curve. Both the crude oil prices and P_{NE} levels are required to assess when ethanol production occurs without ethanol policies.

Having established the intercept of the ethanol supply curve to be P_{NE}, we are now in a position to establish the redundancy in the ethanol price premium due to ethanol policies, that is, "water" in the ethanol price premium. In Figure 3.1, it is the difference between P_{NE} and the free-market price of ethanol P_{E0b} [defined by equation (2.2) and converted into dollars per bushel by equation (2.1)]. The only impact of ethanol policy (tax credit or mandates)[3] is the distance P_{Eb} ($=P_C$) minus P_{NE}. Empirical results are given in the next section. We first develop the properties of the ethanol supply curve.

At a corn price above P_{NE}, there is an excess supply of yellow corn to be used as an input for ethanol production. We assume yellow corn and DDGS are perfect substitutes in the figure (i.e., the relative price of DDGS and corn is equal to unity); the demand curve D_{NE} is for both yellow corn and DDGS for prices greater than P_{NE}.

The amount of yellow corn produced at price P_C is Q_C and the amount to be consumed (in non-ethanol industries) is C_{NE}.[4] Thus, for any price P_C—linked to the ethanol price—the horizontal difference between S_C and D_{NE} in the first panel of Figure 3.1 represents an amount of yellow corn removed for ethanol production. Multiplying this quantity by the parameter $\beta = 2.8$, we obtain a corresponding ethanol supply curve constructed under the assumption that no ethanol co-product exists (not shown in Figure 3.1). Note that the intercept of this curve adjusted for units, corresponds to P_{NE}. However, there inevitably is a co-product of ethanol production, and it needs to be taken into consideration when modeling the corn market.

The assumed perfect substitution of DDGS and yellow corn in Figure 3.1 implies a one-to-one replacement of yellow corn—that would otherwise be consumed as a feed—with the ethanol co-product.[5] Because of this *recycling effect* of the ethanol co-product, additional yellow corn is available for ethanol production. This process continues until the marginal increment in yellow corn that could be used for ethanol is zero. One initial bushel of corn corresponds to $1/(1 - \gamma) = 1.44$ bushels of yellow corn processed for ethanol in the end. By definition, the size of the recycling effect is equal to the total amount of the co-product that is, 0.44 bushels for each one initial bushel of corn.

The implicit demand curve for yellow corn D_{NEY} in the first panel of Figure 3.1 is derived by horizontally subtracting the amount of the co-product from D_{NE} at any corn price above P_{NE}. By construction, D_{NEY} is flatter than to D_{NE}.[6] The elasticity of ethanol supply is empirically found to be on average 2.7 times bigger if one adjusts for the recycling effect.

Ethanol Price Premiums due to Biofuel Policies

Given the two alternative policies under consideration (mandates versus tax credits, if any) have such differing implications for the ethanol market price, how can we identify which one is the binding biofuel policy? One way to determine the binding policy is to compare the observed ethanol market price with the one predicted by equation (2.3). If the difference between these two prices is close to zero, it is very likely that the tax credit is binding. However, if the difference is positive, it implies that the mandate is binding. We designate the difference between the observed ethanol price and the price with a tax credit equation (2.3) as the "mandate premium over the tax credit." However, the total ethanol price premium due to ethanol policy is defined as the difference between the observed ethanol price and the free-market ethanol price defined by equation (2.2).

We use different studies to get estimates. Using monthly Iowa data [specific to data on an ethanol plant coming from the Center for Agricultural and Rural Development (CARD) of the Iowa State University], Figure 3.2 plots each of these two policy price premiums (in dollars per bushel, using the calculated positive/negative profits in ethanol production). The estimates of the ethanol price premiums are adjusted (see below) for "water"—we use annual values calculated by Drabik (2011). We also adjust the data for predicted ethanol (corn) prices that go above the actual values (because of blending constraints and other reasons). The solid line in Figure 3.2 includes negative values (unlike for the dotted curve) to show how premiums went negative in the earlier time periods due to ethanol-blending constraints at the consumer level.

The total price premium (excluding negative values) averaged $1.23 per bushel for the period September 2007 to August 2011 ($1.41 per bushel

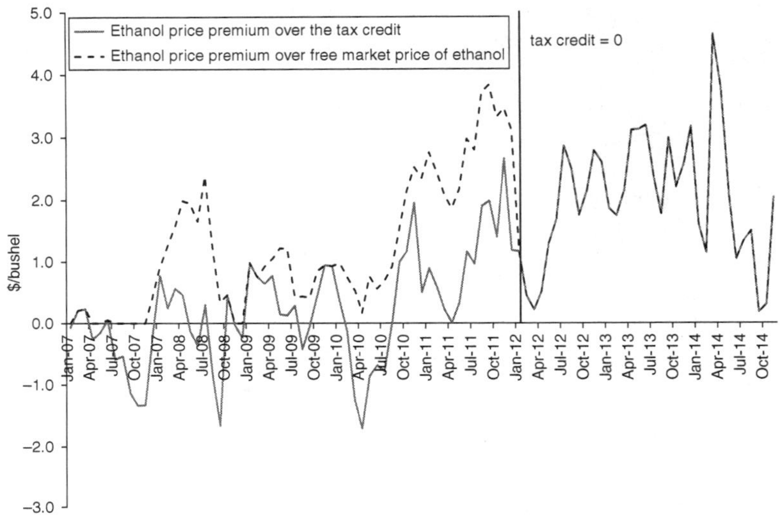

Figure 3.2 Ethanol price premiums.

Note: Data excludes negative values, "water," and when the predicted corn prices exeed actual values

Source: Calculated using data from CARD (2014a), AMRC (2014), and EIA (2014a).

for the time period January 2007 to July 2014). The mandate premium over the tax credit averaged \$0.42 and \$0.92 per bushel for the same two time periods, respectively.

Using annual data of average US farm prices for calendar years 2008–2011, Drabik (2011) obtains similar second to last row in Table 3.1. The first row gives the free-market ethanol price derived from equation (2.2) but in dollars per bushel.[7] As expected, it is very low and the next two rows in the table show the non-ethanol price (P_{NE}) and actual corn price (P_C) are much higher. The ethanol price premium due to policy is very high as shown in the fourth row in Table 3.1 but after adjusting for "water," the net ethanol price premium declines to \$1.16 per bushel. The reasons for this include the de facto MTBE ban up to 2008; the binding US mandate in 2008–2009; and US ethanol prices linked to high world prices. Note that the United States was the biggest ethanol exporter in the world for the period 2010–2012 (for more details see Box 1.3).

Data in the last row of Table 3.1 shows how important it is to adjust the data for P_{NE} and "water" as the latter exceeds the net ethanol price premiums. Otherwise, the price impact attributed to ethanol policies would be overstated by more than 100 percent.

Table 3.1 Estimating "water" in the ethanol price premium due to ethanol policies ($/bushel)

	2008	*2009*	*2010*	*2011*
Free-market price of ethanol (P_{E0})	1.27	0.69	1.33	2.55
Non-ethanol price of corn (P_{NE})	3.59	2.70	2.67	4.11
Observed price of corn (P_C)	4.78	3.75	3.83	6.01
Ethanol price premium due to policy	3.51	3.06	2.50	3.46
"Water"	2.32	2.01	1.34	1.54
Net ethanol price premiums	1.37	1.18	0.57	1.52
"Water" as a % of price premium	66%	66%	54%	45%

Source: Drabik (2011).

The Impact of Biofuel Policies on Grain/Oilseed Prices

To answer the key questions of this chapter, namely how much of the food grain and oilseed price increase is due to biofuel policies, we explore two counterfactuals: the reality of the new corn–ethanol price link; and the case of no biofuel policies (and, thus, no corn–ethanol price link). We analyze each in turn.

The Counterfactual of a Corn–Ethanol Price Link

To analyze how much of the price increase would have occurred irrespective of other factors, we estimate the lowest corn prices could have gone when locked onto crude oil prices through ethanol (gasoline) prices and the tax credit. We use equations (2.1) and (2.3). The corn price must be equal to or higher than this level. As we showed in Figures 2.3 and 2.4 (and in the discussion of Table 3.1), we need to adjust for "water" in the ethanol price premium. We also make adjustments to account for predicted prices exceeding actual prices. This will also allow us to obtain a more accurate picture of the net impact of biofuel policies on crop prices compared to the new counterfactual of the corn–ethanol price link. For ease of exposition, we use the following notation:

P_C: observed corn price

$P_C^{t_c}$: the predicted corn price assuming the tax credit was binding throughout the time period [using equations (2.1) and (2.2) in Chapter 2]

P_C^{lowest}: the lowest corn prices can go, where P_{NE} represents a floor and the predicted corn price with the tax credit is not allowed to be higher than the actual corn price.

In other words,

$$P_C^{\text{lowest}} = P_{NE} \text{ if } P_C^{t_c} < P_{NE}; = P_C \text{ if } P_C^{t_c} > P_C; \text{ otherwise} = P_C^{t_c}.$$

The corn price premium over the tax credit is defined as the difference between the actual corn price and the lowest corn prices can go (in other words, $P_C - P_C^{\text{lowest}}$).[8]

Table 3.2 gives some key results. The average corn price for the 2000–2001 to 2005–2006 is $2.10 per bushel and for the period 2005–2006 to 2011–2012 it is $4.68 per bushel. The increase in corn prices over this time period is thus $2.58 per bushel, or 123 percent. We provide results assuming profits in ethanol production—the preferred method when analyzing historical data series because it represents what actually happened.[9] The lowest corn prices could go in the period 2005–2006 to 2011–2012 averages $4.15 per bushel (if one used zero profits instead, it would be $4.22 per bushel), leaving an average mandate premium over the tax credit to be $0.53 per bushel (=$4.68 − $4.15), or 21 percent of the corn price increase.

However, before attributing this premium to shifts in the supply and demand for corn, we first need to make an adjustment for the situations when the mandate is binding even without such supply and demand shifts. This situation is when the amount by which the premium over the tax credit is higher than the gap between the lowest corn prices can go P_C^{lowest}

Table 3.2 How much of the corn price increase is due to non-biofuel factors?

	Prices	*Value ($/bushel)*
1	Baseline corn price (2000–2001 to 2005–2006)	2.10
2	Average corn price (2006–2007 to 2011–2012)	4.68
3	Increase in corn price	2.58 (123%)
4	The lowest corn prices can go[a]	4.15
5	Premium over the tax credit[b]	
(a)	Gross	0.53 (21% of price increase)
(b)	Deduct when the mandate is binding even without supply/demand shifts[c]	−0.05 (19% of price increase)

Notes:

[a] Average predicted corn price defined by equation (2.2), adjusted to disallow for observations (1) exceeding actual corn prices; and (2b) falling below the estimated non-ethanol corn price P_{NE}.

[b] The premium over the tax credit is defined as the observed corn price minus the lowest corn prices can go with the tax credit (note that the tax credit expired at the end of calendar year 2011).

[c] Accounts for the amount by which the premium over the tax credit is higher than the gap between the baseline corn price and the lowest corn prices can go with the tax credit.

Source: de Gorter and Drabik (2014).

and the baseline corn price (average corn price of \$2.15 per bushel before the advent of ethanol).

To calculate this adjustment factor, we proceed in three steps. First, we calculate the difference between the minimum corn prices and the baseline price. Denote this distance as *a*. Then we calculate the difference between the premium over the tax credit and the distance *a*. We then sum all positive resultant differences over time. Finally, we divide the sum by the total number of observations in the sample. The obtained adjustment factor is \$0.05 per bushel, meaning that the net share of the corn price increase due to the mandate premium over the tax credit falls to 19 percent (last row in Table 3.2).

To put this last figure into perspective and to highlight the importance of our results, let us consider the results of Roberts and Schlenker (2013) who, like most papers analyzing the impact of biofuels on grain and oilseed markets, do not model the biofuel policy or the crop-biofuel price links. Their approach considers biofuels as simply a shift in the demand for corn. The authors conclude that grain and oilseed prices increased 30 percent in the 2010–2011 crop year due to corn–ethanol. If one adjusts for the fact that 24 percent[10] of the corn processed into ethanol was recycled as (equivalent) corn, then 22.7 percent of the grain and oilseed price increase is due to biofuels, leaving 77.3 percent to supply and demand shifts. However, we showed earlier that by recognizing the new reality of the corn–ethanol–gasoline price links, the effect of supply and demand shifts on corn prices in 2010–2011 was approximately 20 percent overall. The change in consumer surplus due to biofuels in Roberts and Schlenker (2013), adjusted for our more accurate recycling estimate, is a loss of \$136.2 billion. But the loss increases to \$410 billion (more than three times the loss estimated by Roberts and Schlenker (2013))[11] if one considers the new corn price counterfactual.

Furthermore, what are the implications of our analysis above for Mitchell's (2008, 17) controversial and widely cited paper in which he concluded: *"Most of the remaining 70–75 percent increase in food commodities prices was due to biofuels and the related consequences of low grain stocks, large land use shifts, speculative activity and export bans"*? We note that, in cases where the corn prices were the lowest they could be, *"The related consequences of low grain stocks, large land use shifts, speculative activity and export bans"* are probably immaterial and could have only impacted the mandate premium over the tax credit. We showed that this mandate premium was a small share of the total corn price increase. So was Mitchell's (2008) attribution of biofuels policies' impact on grain and oilseed prices alone severely underestimated, although the profession ridiculed his analysis as ignoring the "perfect storm" and as being too simplistic? Our analysis suggests so. In fact, we will argue in the next chapter that the 2008 corn price increase (and by extension, all food grain and oilseed price increases, given the law of one relative international price for grains and oilseeds) was basically entirely due to biofuel policies as corn and soybean prices were locked onto crude oil prices all the way up and all the way down (free-market ethanol prices would have been zero without policies as the market could never have reacted to produce significant amounts of ethanol and cause the crop–biofuel price link in 2008).

The Counterfactual of No Biofuels

Although not a reality anymore, it is standard academic practice to estimate what the grain/oilseed prices would have been if there had been no biofuels and no corn–ethanol price link. We are also in a position to determine the impact ethanol policy has on corn prices compared to no biofuels. A summary of the data is given in Table 3.3 for the period 2008–2011. All data are converted into dollars per bushel. We compare Drabik (2011) to Cui et al. (2011) for the one-year data available in the latter study.

The percentage change in corn prices due to ethanol policy adjusted for "water" in the ethanol price premium ["water" is zero in Cui et al. (2011)] averages 40 percent over the four years in Drabik (2011) (53 percent for Cui et al. 2011, for the year 2009). If one ignores "water" as Babcock (2013) suggests, then the average corn price increase due to ethanol policy skyrockets to 261 percent. Cui et al. (2011) are able to get no "water" by calibrating their baseline for 2009 to have the tax credit binding with the gasoline price by 36 cents per gallon higher than its observed value for 2009. If one applies an ethanol price premium over free market levels adjusted for these 36 cents per gallon and maintaining the assumption of no "water," corn prices would be predicted to increase 93 percent due to ethanol policy (see also Box 3.1).

Table 3.3 Percentage change in corn prices due to ethanol policies (compared to no biofuels)

	Drabik (2011)				*Cui et al. (2011)*
	2008	*2009*	*2010*	*2011*	*2009*
Adjusted for "water"	33	39	43	46	53[a]
Ignore "water"	270	451	188	135	93[b]

Notes:

[a] Cui et al. (2011) assume exactly zero water in the tax credit.

[b] Percentage increase in corn price if Cui et al. (2011) reduce gasoline price 36 cents per gallon to match the observed gasoline price.

Source: Drabik (2011), Cui et al. (2011).

Box 3.1 Did US ethanol policy reduce corn prices instead?

The results of some economic analyses imply that the decrease in gasoline prices due to ethanol production is higher than the increase in ethanol prices due to ethanol policies as we have measured it so far! Although such a finding sounds surprising, Du and Hayes (2012) actually conclude that, if there had been no ethanol production, the gasoline prices would have increased $0.89 and $1.09 per gallon in 2010 and 2011, respectively. Such an increase in gasoline prices is far higher than the decrease in ethanol prices (each translated into a dollars per bushel change in corn prices) we have calculated if there had not been any ethanol policy and, thus, no ethanol production in these two years. Therefore, according to Du and Hayes's analysis, ethanol policy caused a net decrease in corn prices!

We summarize the data in the following table.

The change in corn prices with no ethanol production: our approach versus Du and Hayes (2012)

The Du and Hayes's (2012) estimates of the gasoline prices increase due to the elimination of the ethanol production are converted to an ethanol price equivalent by simply multiplying the change in gasoline prices by 0.70 (the miles per gallon of ethanol relative to gasoline). To obtain the equivalent change in corn prices in dollars per bushel,

Table B3.1

	2010	*2011*
Du and Hayes (2012)		
Change in gasoline price if no ethanol		
$/gallon	0.89	1.09
$/ethanol-equivalent gallon	0.62	0.76
$/bushel corn equivalent	2.49	3.05
Our approach		
Change in ethanol price premium if no policy and, therefore, no ethanol[a]		
$/gallon	−0.72	−0.29
$/gasoline-equivalent gallon	−1.03	−0.41
$/bushel corn equivalent	−1.16	−2.88

[a] Takes into account a change in gasoline prices through changes in the crude oil price due to ethanol production in the first place (see Drabik, 2011 for details).

simply multiply the change in ethanol prices by four, the average corn–ethanol price multiplier over the period.

Under our approach summarized in this chapter, we first show the predicted change in ethanol prices due to the removal of ethanol policies. We then convert this into an equivalent change in gasoline prices by dividing by 0.70. The final row gives the decline in dollars per bushel corn equivalent.

Now reverse the process. Assume there was no ethanol policy in the first place and then the observed ethanol production comes online in these two years. Du and Hayes (2012) predict a decline in corn prices by more than we predict corn prices would increase!

How did Du and Hayes (2012) get such a huge decline in gasoline prices due to ethanol production, so much that it swamps the effect of our calculated increase in ethanol prices due to ethanol policy? They claim ethanol production generated excess production in oil refining, causing processing margins to plummet (yet refining margins only averaged 27.8 cents per gallon historically). Elam (2012) demonstrates the flaws in their methodology and using a more complete model, he shows the increase in gasoline prices would have never happened. Besides, the United States could export less gasoline. Knittel and Smith (2014) argue the estimates by Du and Hayes (2012) were generated by implausible economic assumptions and spurious statistical correlations.

Drabik (2011)'s estimates show that US ethanol policies increased the corn prices between 33 percent and 46 percent in the period 2008–2011. However, these values are underestimated because, although the analysis incorporates the effect of a change in inventories, it does not recognize that the initial level of corn prices are determined where supply intersects the total demand for current consumption and stocks—a demand curve much more elastic than the demand for current consumption only. Stocks are now lower and a new regime exists between stockholding and market prices. The price response to a supply or demand shock depends on the level of inventory (Hochman et al. 2011; Wright 2011, 2014; Carter et al. 2013) and, therefore, if inventory is drawn down—itself because of biofuels as an unexpected event—then, any shock in corn supply and demand results in a higher change in corn prices than if there were no biofuel policy. This also has implications for price volatility discussed in a later chapter.

The Other $1 Billion Per Day Farm Subsidy

Each year, the OECD publishes the "total subsidy equivalent" for agriculture. It averaged $375.3 billion for the past six years (OECD 2013). The argument that farm subsidies in rich countries contribute to global poverty by distorting prices has become an article of faith among many advocates for free trade and global justice. This has become, over the years, the "$1 billion dollar a day farm subsidy" popularized by, among others, the President of the World Bank, Oxfam, the Food and Agriculture Organization of the United Nations (FAO), and many other nongovernmental organizations (NGOs) and international organizations (Takarick 2008; Peterson 2009).[12] This formed the basis for developing countries demanding significant trade liberalization in agriculture and was the major reason for the demise of the World Trade Organization's (WTO) Doha Development Round trade negotiations. There was much ado about something very significant—and rightly so.

Activist economists in the World Bank went even further and calculated numbers like:

> The World Bank reports (September 25, 2002) that a study of farm subsidies in the European Community has shown that the annual subsidy paid to dairy farmers would pay for a round-the-world air trip for every one of the 21-million cows in the Union. This would be a first-class trip, with stops in London, Hong Kong, Shanghai, Hanoi, Siem Riep, Brisbane, Los Angeles and San Francisco, where three-day breaks in each destination, in five-star hotels would enable discussions with local cows about comparative dairy policy and conditions ...[13]

The chief economist of the World Bank at the time had this to say:

> While most Africans are barely surviving—or often not surviving—on less than US $2 per day, the average European cow is pampered with US $2.50 a day in subsidies, Nicholas Stern, the World Bank's chief economist, pointed out during a speech he gave last November at the Center for Economic Studies in Munich, Germany.[14]

James D. Wolfensohn, president of the World Bank, accused

> Wealthy countries of "squandering" $1 billion a day on farm subsidies that often have devastating effects on farmers in Latin America and Africa.[15]

Lower prices cause poverty was the claim. But, Swinnen and Squicciarini (2012) ask why did international organizations (e.g., the World Bank and

FAO) and NGOs (e.g., Oxfam), all of a sudden, criticize biofuel policies of OECD countries' biofuel policies (that instigated the food price rise in 2007–2008) when, previously, these organizations criticized the same countries' agricultural subsidies and trade policies for depressing world agricultural prices? In other words, if low prices were bad before, why are high prices all of a sudden a crisis now (Swinnen and Squicciarini 2012)?

Nevertheless, since September 2006, we determine that there has been another "$1 billion dollar a day farm subsidy." We estimate the transfers from food grain and oilseed consumers to crop producers by subtracting the baseline corn price ($2.10 per bushel for the period 2001–2005 as reported in Table 3.2) from the lowest prices corn could have gone on average over the ensuing six years when locked onto the ethanol and, thus, gasoline (crude oil) prices (adjusting for "water" and that predicted corn prices could not exceed observed corn prices). We use profits in ethanol production as this is the more realistic way to analyze historical "what if" questions. Using the lowest corn price when locked onto ethanol and gasoline (crude oil) prices, we invoke the law of one relative international price of food grains and oilseeds.[16] All other grains and oilseed prices are assumed to maintain their historical value relative to corn [as in the work of Roberts and Schlenker (2013) and Wright (2014)]. Empirical evidence that relative prices of food grains and oilseeds have not changed after September 2006 is given in de Gorter et al. (2013a) (see also Box 1.3). This is because all grains and oilseeds compete for the same land (are substitutes in production) and are highly substitutable in consumption. If the price ratios fall outside of a reasonable range, farmers could take profits by switching from the lower relative price commodity to the higher until the ratio is restored (likewise for consumers to save money).

Consistent with the law of one relative international price, we transform the change in the corn price to equivalent changes in the price of wheat, rice, and soybean prices by multiplying the change in the corn price per metric ton by the relative historical ratios of the wheat, rice, and soybean price to the corn price (1.28, 2.44, and 2.33, respectively).

Table 3.4 provides an estimate for each year, commodity group, and world totals. The annual average transfer over the six years reported in Table 3.4 is $328.3 billion (compared to the average "total farm subsidy equivalent" reported by the OECD of $375.3 billion). This amounts to approximately seven times the total annual foreign aid spent on developing countries by rich countries. The data in Table 3.4 does not adjust for the fact that producers were insulated from international prices; especially in 2008 (see Chapter 6 for a detailed discussion). Thus, the estimates are overestimated in Table 3.4, especially for the year 2008.[17] On the other hand, data in Table 3.4 are underestimates for several reasons. If one includes the

Table 3.4 Transfers to grain/oilseed producers because of biofuel policies (million dollars)[a]

	Coarse grains	*Wheat*	*Rice*	*Oilseeds*	*World total*
2006–2007	46,922	36,115	48,803	44,670	176,510
2007–2008	111,544	80,756	109,783	94,460	396,543
2008–2009	48,475	38,107	47,875	40,319	174,776
2009–2010	50,891	39,912	49,133	47,022	186,959
2010–2011	143,561	108,814	143,579	138,688	534,642
2011–2012	136,690	105,666	135,217	123,367	500,941
Average	**328,395**				

Note: [a] The analysis looks at US corn–ethanol market only and, therefore, requires the "law of one international relative price" to hold in order to extend to all food grain-oilseed price changes.

Source: Calculated using data from Drabik (2011), Table 3.2, and World Agricultural Supply and Demand Estimates Reports (WASDE) (various issues).

profits of oil extracted from DDGS that is now occurring in most ethanol plants, annual transfers reported in Table 3.4 would, on average, increase by $58 billion. It also excludes taxpayer transfers that were over $10 billion per year (Steenblik 2007) and the price-increase effects on other crops not analyzed, although the grains/oilseeds analyzed here account for the bulk of calories consumed by mankind directly or indirectly through meat and dairy products. It should be noted that the $1 billion per day subsidy due to biofuel policies represents the minimum amount of transfers from crop consumers to producers that would have occurred regardless of supply/demand shifts or changes in inventory in food commodity markets.

Are We at Free-Market Corn Prices in 2014?

In late June of 2013, a news report on ethanol prices stated:[18]

> Sharp losses developed in the front-month ethanol futures contract. With corn prices falling below $5 per bushel in the futures market, traders are quickly starting to prepare for a time when high corn prices may not dictate the direction of the ethanol market.

To translate this passage, recall that one of the two possible states of nature is when the ethanol (and, thus, corn) prices float up and away from the gasoline (and, thus, crude oil) prices. This occurs when mandates are binding in which case "corn prices dictate the direction of the ethanol market" as the news report suggests (we showed this in Chapter 2 with our theory of blend mandates). Our data from June 2013 to August 2014 shows that mandate premiums have prevailed, although predicted corn prices with zero profits in ethanol production are equal to actual corn prices. However, excess profits have offset mandate premiums to generate such an outcome. In the quote earlier, the industry was "preparing" for the possibility that corn prices would equal to the lowest corn prices can go with profits. In other words, if after June 2013 mandate premiums were zero (along with profits), then the other state of nature would be in play, namely, corn prices are locked onto crude oil prices (through ethanol and gasoline prices) and, thus, corn prices would strictly follow ethanol prices (as the passage above shows the industry was preparing for such an outcome). The data for September 2014 would suggest such a scenario.

We have been in the "mandate" state of nature for some time but, in mid-2013, traders apparently were thinking a switch to the other state of nature was eminent. The other state of nature is when the corn price is at its *lowest* possible level when *locked* onto the crude oil price through the ethanol and gasoline price links (plus the tax credit when it existed)—this means corn prices would follow ethanol prices.

Did the corn price reach rock bottom in 2013–2014? The observations for November 2013 to September 2014 in Figures 2.3 and 2.4 suggest so but recall in our discussion of the prediction error in Chapter 2; profits were at an all-time high in this recent time period. Therefore, it begs the question of what would actual corn prices be if profits, instead, were zero (or at "normal" levels). Recall we concluded profits were so high because of bottlenecks in ethanol supply and record high levels of the price of DDGS relative to corn (4.82 in December of 2013).[19]

We know from our theory that, when a mandate is binding, the incidence of removing an exogenous wedge between corn and ethanol prices (in the form of eliminating excess profits) will have ethanol prices falling much more than corn prices rising (the opposite if the market is locked onto gasoline prices). So, in that regard, we expect only modest increases in the corn price. Thus, we predict corn prices were at least close to rock bottom in these months.

However, the high value of DDGS prices contributes to the high profits independently of profits due to the supply bottleneck. The issue then becomes whether there are ethanol production constraints that prevent expansion in ethanol production to eliminate profits by lowering the relative price of DDGS. Without constraints, the relative price of DDGS would decline and corn prices increase, such that corn prices would not have been at rock bottom in this time period. This issue requires far more empirical analysis in order to ascertain how close corn prices were to the lowest they can go in this time period and how much of the profits are due to relatively high DDGS prices as opposed to supply bottlenecks. Relaxing supply bottlenecks will reduce ethanol prices and increase corn prices, ceteris paribus. In a situation where relatively high DDGS prices prevail together with ethanol production-capacity constraints, profits can only be eliminated through (sufficiently) expanding production capacity. In the process, the corn price increases, thus lowering the relative price of DDGS and further reducing the profits.[20]

Is the Tax Credit Really Redundant?

We showed in Figures 2.3 and 2.4 that, from November 2013 to September 2014, the predicted corn prices under free markets (lowest when locked onto crude oil prices, assuming no excess profits in ethanol production) were very close to observed corn prices. However, ethanol price premiums due to the mandate and strong world prices, among other things, have prevailed for all of these observations except the very last one. The reason for actual corn prices being about equal to predicted free-market levels is the excess profits in ethanol production. But what if these disappear, along with ethanol price premiums? Corn prices would be very close to free-market levels, given our theory of corn–ethanol price links and ethanol price premiums.[21] This also means "water" is zero (the non-ethanol price premium has declined with shifts in corn supply and demand curves) and corn prices could, therefore, be locked onto crude oil prices.

Under such a potential scenario, a tax credit would increase corn prices by the full amount of the tax credit. We estimate this to be $2.31 per bushel for the November 2013 to July 2014 time period. This is because the corn–ethanol price multiplier averaged 4.53 over this period[22] and the tax credit was 51 cents per gallon [a federal tax credit of 45 cents per gallon and state tax credit of 4.8 cents per gallon as reported in the well-publicized research of Koplow (2009)]. Multiplying 4.53 by $0.51 per gallon generates the $2.31 per bushel estimated corn price increase if the tax credit was not allowed to expire. [23] Furthermore, by invoking the law of one relative international price for grains and oilseeds, consumer transfers to grains and oilseed producers would have amounted to $445 billion (using average prices in the seven months between September 2013 and March 2014 and prorating to an annual basis for projected world grain/oilseed production in 2013–2014).[24]

Therefore, allowing the US tax credit (for corn and sugarcane ethanol) to expire may be an important development in lowering grain/oilseed prices. Many economists argue the tax credit is redundant (Babcock 2011; Tyner 2013),[25] so that eliminating the tax credit should have no impact. Smith (2012) in a paper entitled, *Giving an Inch and Keeping a Mile: Why the Corn Lobby Let the Ethanol Tax Credit Expire*, argues the tax credit makes little difference; the mandate is the true source of support to ethanol prices.

This may be true when the ethanol price premium due to the mandate is greater than the tax credit (when it surely was for much of the time period, especially leading into 2009 and with the drought in 2012–2013), but otherwise the premium is lower than the tax credit and can even reach zero. In these situations, the tax credit could make a difference.

Thus, by taking the new counterfactual and employing the methodology developed so far in this book, we assess the conventional wisdom that a tax credit is redundant and come to a very different conclusion. It depends on the ethanol price premium due to the mandate over the tax credit (often equal to zero) and more importantly, in the near future, depending on grain/oilseed supply/demand conditions. Maybe the corn–ethanol lobbyists have given a mile (445 billion dollars' worth!) and kept an inch.

However, we cannot know for sure what would have been the outcome had the market been restored to equilibrium of zero profits (in other words, we do not know if corn prices really were at the lowest they can go in the time period under consideration—this has yet to be determined)—nevertheless, this is exactly how one has to assess the situation. We have been in a new era and the profession analyzing the effects of biofuel policies has to recognize that first and foremost.

Concluding Remarks

This chapter covers a lot of important topics. Building on the theory of price links in the previous chapter, we determine the non-ethanol corn price (the intercept of the ethanol supply curve) and subtract the level of "water" in the ethanol price premiums due to ethanol policies (otherwise, the impact of ethanol policies on corn prices would be overestimated). We find empirically that more than half of the ethanol price premium over free-market levels is "water" or redundant. Nevertheless, we find that 80 percent of the corn price increase in the period 2007–2011 would have occurred due to ethanol policies, regardless of supply and demand shocks and all other "perfect storm" factors, because of the counterfactual of the corn–ethanol price link, with biofuel policies putting a variable price floor under corn prices. The only way supply or demand shifts in the corn sector could have affected corn prices directly was if there were an ethanol price premium above that implied by the tax credit (the only other possible state of nature). All other factors like supply and demand shifts can only affect the mandate premium above the tax credit, representing a mere 21 percent of the observed corn price increase. This is a more sober and realistic result than most of the literature would have had you believe in the past eight years. Most economists have thoughtfully analyzed the large upswing in prices assuming business as usual. It is only true insofar as one fails to recognize the fundamental change in the function of the market that took place in 2006–2007 when suddenly crop, biofuel, and energy markets became linked.

We determine that almost 80 percent of the increase in crop prices would have occurred, regardless of all other factors, due to biofuel policies alone. This is the more relevant counterfactual, given the new reality of crop–biofuel price links. Because the law of one relative international price of food grains and oilseeds holds (because of competition for land and substitution in demand), 80 percent of food grain and oilseed price increases are due to biofuel policies. When the tax credit is binding (when it exists; otherwise "free markets"), supply/demand shifts cannot significantly affect the corn price. Because the average mandate price premium over the tax credit represents 20 percent of the corn price increase, the remaining 80 percent must be due to the price link caused by biofuel policies and the tax credit alone (unless some ethanol would have been produced under free markets but the link between corn and ethanol prices would never have occurred without energy and environmental policies; therefore, we postulate all ethanol production is due to ethanol policy through 2008). Many academics and NGOs think a 20–30 percent increase would be dramatic (e.g., Roberts and Schlenker 2013; Oxfam 2008).

We then estimated what would have happened if biofuel policies had not been in place. One may think this to be the important counterfactual to examine, although difficult to address with any degree of accuracy. Although these good old days do not exist anymore, this counterfactual is also an interesting and important exercise, albeit purely academic at this point. We estimated that corn–ethanol increased crop prices 40 percent on average. For various reasons, this may be a substantial underestimate of the true impact (in particular, we do not take into account the nonlinear impact on prices of low stock levels as emphasized, among others, by Wright 2011, 2014).

We estimate how low corn prices could have gone and, by extension, given the law of one relative international price of grains and oilseeds, all other crop prices, too. The proper counterfactual must recognize that the lowest possible corn price occurs when ethanol (and, thus, corn) prices are locked onto gasoline (and, thus, crude oil) prices through the tax credit. We then estimated the transfers from grain/oilseed consumers to producers and found another $1 billion per day "farm subsidy," in addition to that estimated annually by the OECD (OECD 2013). Thus, biofuel policies represent the "other $1 billion per day farm subsidy." The controversy over farm subsidies was the underlying reason for the failure of the WTO's Doha Development Agenda trade negotiations. The $1 billion per day subsidy due to biofuel policies seems both bewildering and expensive, and equally controversial. Nonetheless, this transfer represents the minimum amount of transfers from crop consumers to producers that would have occurred regardless of supply/demand shifts as long as biofuel policies were in place.

Similar annual potential cost savings to consumers and taxpayers were estimated by allowing the US tax credit to expire, given that corn prices may in the near future become locked onto crude oil prices (through and ethanol and gasoline prices) with no excess profits in ethanol production and no mandate premiums. The average corn price in this time period was $4.39 per bushel and our prediction on how low corn prices could go is $4.58 per bushel if there were no profits (no capacity constraints) or ethanol price premiums. Under these conditions, the tax credit could have added approximately $2.31 per bushel to the price of corn had it not been allowed to expire.

CHAPTER 4

A FORENSIC ANALYSIS OF THE FOOD COMMODITY PRICE BOOM OF 2008

This chapter will contribute to the understanding of how this price spiral happened, which has quickly become one of the most analyzed period of food commodity price movements in history. As Timmer and Dawe (2010, 8) state:

> No lessons from the food crisis are of much relevance without understanding how this price spiral happened.

Our analysis is unique as we recognize that biofuel policies caused a link between corn (soybean) and ethanol (biodiesel) prices beginning in September 2006. This meant that near-futures prices of corn and soybeans were locked onto crude oil prices all the way up and down the peak in 2008 (see Figure 4.1). In other words, the price of feed grains and oilseeds would climb in 2008 regardless of supply and demand shifts (e.g., demand growth, bad weather) or the *quantity* of crops diverted to biofuels. One can even rule out speculation as a cause of the coarse grain and oilseed price increases. However, through the law of one relative international price of grains and oilseeds, all crop prices moved together. Biofuel–crop price links was the engine that pulled all prices up (and down).

Although the blenders' tax credit was only worth approximately $2 per bushel for corn, the entire price spike observed in this period can be attributed to biofuel policies because the price link would not have been created without policy. No free-market ethanol production would have occurred in 2008 without this policy although "water" in the ethanol price premium was negative for some months in 2008 (meaning ethanol would have been briefly profitable even without the subsidy). Because feed grains and oilseed markets had no choice but to follow crude oil prices to the top in 2008,

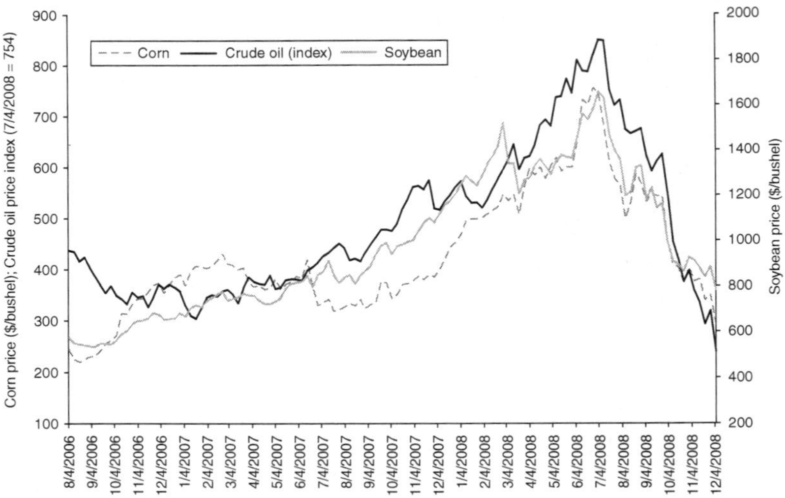

Figure 4.1 Near futures prices for corn, soybeans, and crude oil.
Source: Bloomberg (2014).

this put other crop prices in a corner. Because of the law of one relative international price for grains and oilseeds, these market prices were compelled to go up, regardless of no significant market developments otherwise (or the near-futures market would do it for them, "speculating" that this is what would normally happen anyway).

Nevertheless, wheat and rice prices peaked before corn. In fact, wheat and corn prices even led corn prices part of the way up and down in 2008. Was it indeed a "perfect storm" of supply and demand shocks that led to the price spike? Was it in fact three different perfect storms as some have contended? Or was it all corn and other coarse grains and oilseeds? This relationship between the prices of coarse grains and oilseeds (which had no choice and were locked onto crude oil prices) and that of wheat and rice (which allegedly are not directly linked to other crops, and thereby could plausibly have remained at low price levels) is one of the most understudied aspects of the recent food commodity price spikes.

For 32 months leading up to September 2006, corn prices did not move as both ethanol and crude oil prices skyrocketed (see discussion of Figure 1.4 in Chapter 1). Wheat and corn prices followed each other closely until the last quarter of 2006 when corn prices took off, reacting (rather belatedly) to ethanol production capacity and capacity under construction (see our Figure 1.5 in Chapter 1). This added ethanol production capacity is what allowed the link between corn prices and ethanol prices to form. Although the Central Illinois corn price increased 88 percent from August 2006 to February 2007 (107 percent for No. 2 white corn, Kansas City), the price increase was lower than might have occurred as stocks were being drawn down significantly (Mitchell 2008; Hochman et al. 2011).

When ethanol production capacity was about to double in 2006, corn prices finally reacted two-thirds of the way through the year by basically doubling over the next five months. The Mexican tortilla crisis in January 2007 and the India wheat import ban in February of 2007 were the initial political responses. These were the first of many developing country policy responses to come. This got the ball rolling in terms of establishing the corn–ethanol price link[1] while putting other food grain prices on notice. Approximately 62 percent of the world's land area is devoted to coarse grain and oilseed production—their prices locked onto crude oil prices. Only crude oil price levels and time would tell how things would proceed from here to there. But the table was set.

The 2007–2008 Grains/Oilseed Price Boom: Was it All due to Biofuels?

Our forensic analysis leads us to June 2007 when the wheat price began its journey up, following a slightly downward trend for the previous eight months (a time period within which corn prices almost doubled). Although world wheat production was down 3.9 percent in the 2007–2008 crop year, overall grain production was essentially unchanged that year. Instead, wheat prices were pulled up by this sudden increase in corn prices (Figure 4.2). Corn returned to its upward journey two months later in August 2007 and reached $3.84 per bushel. But our model (outlined in Chapter 2) predicted corn prices should have been $5.14 per bushel (see the dotted line segment in Figure 4.2), given the levels of ethanol, gasoline, and crude oil prices and zero profits in ethanol production. However, ethanol capacity constraints prevented it from reaching such levels. But wheat prices kept going up. Wheat prices paused for three months (September to November of 2007) when corn prices kept marching up—albeit at a slower rate than what it should have been (for the period November 2007 to June 2008, compare the dotted and solid lines in Figure 4.2).

Glancing at Figure 4.2 over the period September 2006 (when corn prices initially spiked) to the low for both crops in December 2008, it looks something like a repeated two-man relay race with wheat and corn intermittently handing the baton to each other in terms of which price goes up first and the other follows. Comparing predicted versus actual corn prices on the way up in late 2007 to early 2008, corn and wheat would have been side by side, both holding the baton. For the second half of the trip down, corn and wheat

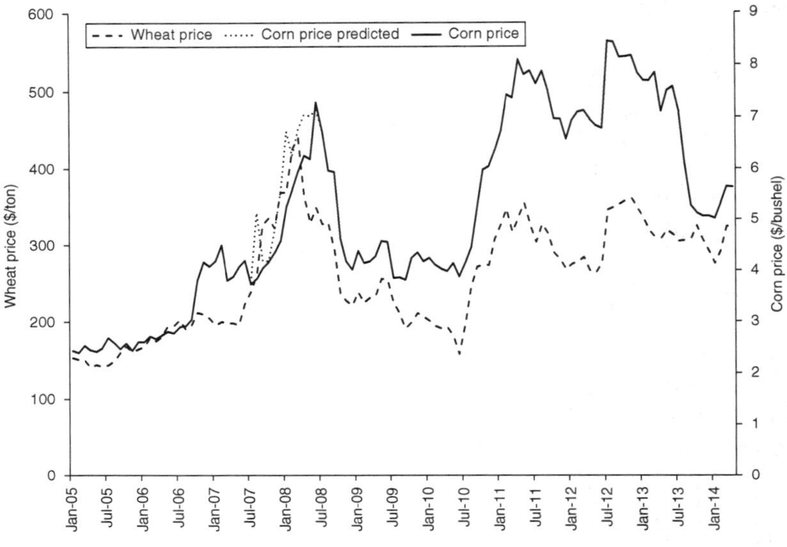

Figure 4.2 Corn–wheat price developments.

Source: USDA Wheat Yearbook Table 20, USDA Feed Grain Yearbook Table 12.

prices were side by side. However, the other hand of corn was sharing the baton side by side with soybeans all of the way up and all of the way down as corn and soybean prices were locked onto crude oil prices (Figure 4.1).

India and Ukraine's export bans on wheat occurred immediately after corn prices finally reacted to high oil prices and skyrocketed to their interim peak in February 2007. Immediately following these two wheat-export bans, the price of wheat (which had risen approximately $3 per bushel since January 2005) actually rose faster than corn prices.

Wheat price increases began their last leg (basically straight up) coinciding with the October 2007 rice-export ban by India, moving from $7 to $11 per bushel in a very short period. Wheat prices peaked three months before corn prices—clearly, wheat prices had overreacted relative to corn and, thus, declined to get back in "equilibrium" and began to decline as corn prices continued to climb. Why did corn prices keep marching up? The reason is corn prices were locked onto crude oil prices, as defined by equations (2.1) and (2.2′) derived in Chapter 2. The proof is given in Figure 4.1 where the near futures prices of corn are locked onto the near-futures prices of crude oil. Figure 4.1 gives the same data for soybeans and crude oil prices (we will provide the soybean–biodiesel price links and multipliers in Chapter 9). Corn and soybeans had no choice: they had to go up as dictated by crude oil prices.

But in May, wheat prices rose with corn, and then both corn and wheat prices went down in tandem all the way to the bottom in December 2008.

What about Rice Prices?

Our focus now turns to rice prices which did not move until the India rice ban in October 2007 (Figure 4.3). Wheat had reached a mini peak and corn had gone up three months in a row. As rice was by far the last price to react, it went ahead of corn at $5.30 per bushel in March 2008 whereas wheat had passed corn earlier in July 2007 after corn had almost doubled to approximately $4 per bushel in February 2007. The moment corn prices reached their interim peak in February 2007, India banned wheat exports. Wheat prices reacted only slightly to the corn price jump. Wheat prices appeared to be somewhat independent, but then accelerated upward with the India wheat export ban and then accelerated even more with the Ukraine ban on wheat exports (with wheat prices overtaking corn prices). And finally, on October 7, 2007, India banned exports of rice, sparking rice prices to increase for the first time. During this time, wheat prices continued their increase on a steep trajectory.

The relative wheat–rice prices in Figure 4.3 shows wheat is far ahead of rice and takes off at an accelerated rate until the India rice export ban.

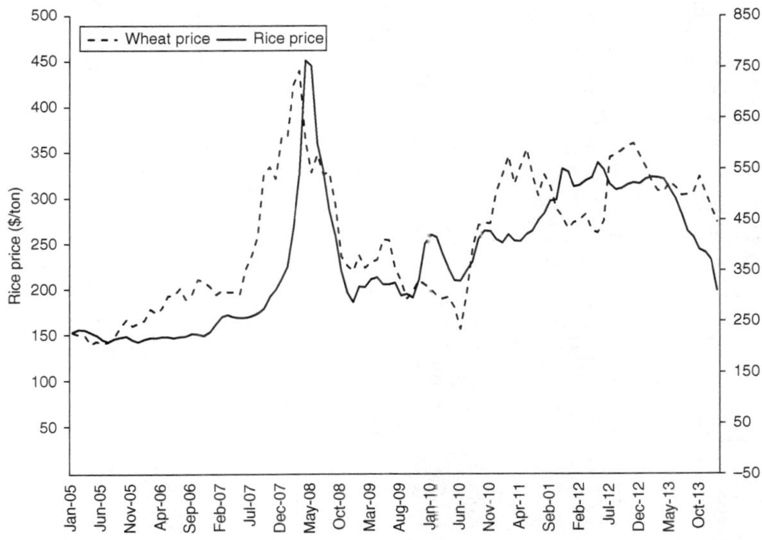

Figure 4.3 Wheat–rice price developments.

Source: USDA Wheat Yearbook Table 20, USDA Rice Yearbook Table 20.

However, on October 7, 2007, India bans rice exports and rice prices began their meteoric rise. Rice prices reached their peak in April and May whereas corn prices kept climbing and peaked one month later in June 2008. Rice collapsed one month before corn but two months after wheat.

There are two possible ways to explain the mercurial commodity prices of 2007. Either three separate markets were bowled over by three distinct sets of supply or demand shocks that all just happened to create price spikes at the same time or the three concurrent price spikes were each caused by the suddenly activated link between grain and oilseed prices and that of crude oil. The prevailing wisdom has relied on the perfect storm theory, which in this case actually supposes three different perfect storms: one in the south Pacific, one in the north Atlantic, and the other in the Indian Ocean. However, this theory (which relies much more on a theory of outliers than Occam's razor) ignores the important links that have always existed between grains and oilseed prices that we speak of as the law of one relative international price. Rice prices went up almost threefold in seven months. What is the explanation for that? An excellent analysis is given in Slayton (2010). There were no major supply or demand shocks in the rice market (Dawe and Slayton 2010, 17):

> Rice market fundamentals were not the cause...the rice crisis was not caused by adverse shocks to rice production or low rice stocks...the world rice to stock ratio was roughly constant in the three years preceding the crisis...world rice trade in first four months of 2008...was 20 percent higher than in the first four months of 2007. The favorable situation as regards production, stocks and trade strongly suggests factors other than basic market fundamentals were at work.

Dawe (2010, vii) in his preface notes:

> [it is] widely acknowledged that the spike in world rice prices had a fundamentally different explanation from the spikes in wheat and maize prices. This while the world rice crisis was undoubtedly shaped to some extent by the same broad events that contributed to price spikes in other world food markets, the world rice economy took on a dynamic of its own, especially in early 2008. In particular, government policies were crucial...in terms of spill-over effects on world markets.

While Timmer and Dawe (2010, 3) note:

> The world rice market ran amok in late 2007 and 2008...[although] "a small decline in overall grain production in 2007...With wheat supplies for near term delivery tight, wheat prices started rising sharply in May 2007...The actual price panic that resulted, however, had little rationale in the [traditional] fundamentals of supply and demand.[2]

We have to agree with these experts' assessment of the situation in rice markets. An easy way out would be to simply conclude that the law of one relative international price for grains and oilseeds was the reason for rice's price ascent—it had to follow feed grain and oilseed prices that were locked onto crude oil prices. Not as much as wheat would have to, given that wheat is highly substitutable with coarse grains and oilseed products in demand and that wheat often competes for the same land. Or did biofuel policies cause the world rice economy to take on a dynamic of its own (Mitchell 2008)? Did the corn price spike ending in February 2007 make governments, state trading agencies, private traders, and even the peasant who hoarded rice nervous and cause them to panic? As the papers by Slayton, Dawe, and Timmer show, there were no untoward market developments in the rice market. Was it even possible that the steep rise in rice prices could be completely unrelated to other markets?

Why did wheat go up faster than corn? Wheat is also used for ethanol in Canada and Europe. So why did wheat prices pause for these three months? Why did wheat prices not follow corn prices from September 2006 to May 2007? Why did wheat prices begin their climb in 2007 two months before corn? Why did wheat prices peak three months before corn in June 2008 when crude oil prices peaked (although corn prices should have been slightly higher than wheat prices as depicted in March), and start going down as corn prices kept going up with crude oil prices? On the face of it, the answers to these questions remain a mystery but, clearly, wheat prices overreacted. Part of the reason corn prices did not go up as early (or as fast) as wheat in the 2008 run-up to the peak were ethanol production-capacity constraints.

We argue wheat prices had to follow coarse grain and oilseed prices that were locked onto crude oil prices (maybe wheat was too as it is used in ethanol production) but in the confusion of biofuels with reports of bad weather and lower stocks as the crop year began plus the totally unexplainable developments in rice prices reality interfered. The literature has offered a number of alternative explanations including the perfect storm of coinciding factors: developing country policy responses like wheat export bans that exacerbated price rises in world markets; bad weather reflected in monthly United States Department of Agriculture (USDA) World Agriculture Supply and Demand Estimates (WASDE) forecasts in world wheat production that turned around by November when forecast production started to increase; low stocks of wheat that could not absorb this bad weather shock; and wheat markets always react earlier than corn because their production season is earlier.

We know crop supply/demand shifts could not have affected feed grain and oilseed prices in 2008: they were locked onto the oil price because of biofuels policies (and underreacted due to ethanol production constraints). However, wheat and rice prices reacted earlier. Was this due to the psychological effects of the developing country policy responses and reduced wheat production? Or was wheat simply catching up to the 2006–2007 corn price boom?

The Bloodbath from the US Financial Crisis: Prices Post 2008

The 2008 US financial crisis, induced by failed regulatory and financial policies, caused the Great Recession. Crude oil prices fell precipitously with corn and soybean prices in hot pursuit. Wheat prices had plummeted three months earlier; rice prices one month earlier. However, all three joined hands after that with corn dropping faster than anticipated by our simulation assuming zero profits (again explained by excess profits in ethanol production due to production constraints). Were these drops because of higher stocks, good weather, and the unwinding of developing country policy responses? Or was the price rise and decline due to the impact of biofuel policies in linking crop prices to crude oil prices where the coarse grain/oilseed sectors had no choice but to go up and down with crude oil, regardless of stock levels and weather developments? Because of the law of one price for grains and oilseeds, all other sectors would have to have followed. How can developing country policy responses affect the corn/oilseed prices when locked onto crude oil prices? Were the developing country policy responses more a response to high prices rather than a cause of it? Or did developing country policy response affect wheat and rice prices significantly? We will address these and other explanations offered by economists in more detail in the literature review in the next chapter.

An Interlude in the Aftermath of the Financial Crisis: December 2008 to June 2010

It is now instructive to analyze the interim period of December 2008 to June 2010. Both corn and wheat ended their straight-down price plummet in December of 2008 and, after June 2010, both corn and wheat went up in tandem—no batons this time around—perhaps reflecting that the market had learned from the prior run up. In this December 2008 to June 2010 time period, it is important to realize that crude oil prices continued to decline but grains and oilseed prices held fast—see Figure 4.4.

We identify four key events in Figure 4.4 leading up to this December 2008 to June 2010 time period. First, crude oil prices led cereal prices early on. Second, the corn wake-up call on ethanol production capacity growth. Third, corn and soybeans were locked onto crude oil prices that increased

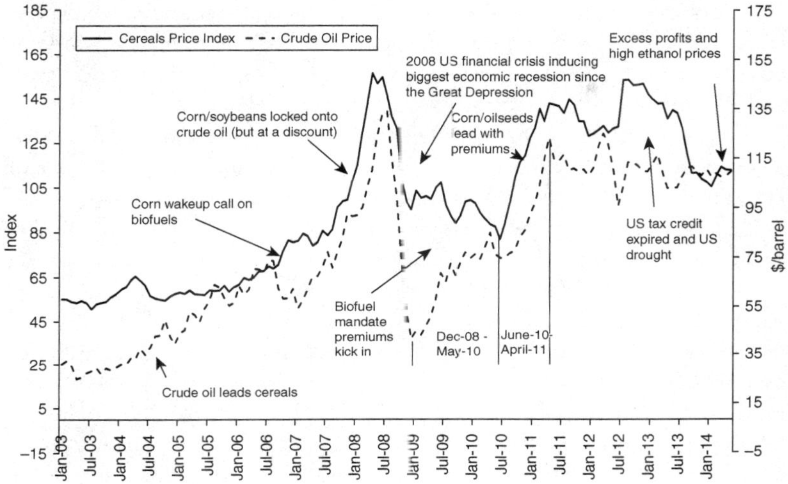

Figure 4.4 Cereal and crude oil price movements.

Source: de Gorter et al. (2013a).

(with wheat and, later, rice prices overreacting). And, fourth, the 2008 US financial crisis was inducing the Great Recession (where the Illinois farm price of corn plummeted to a low of $3.30 per bushel, more than double the $1.62 per bushel low in 2005–2006 when the world was enjoying its biggest economic boom in history!). Cereal prices were jolted by the corn price increase, even if cereal and oil prices went in opposite directions in that time period. However, as cereal prices flattened out in 2007, oil prices continued their steep rise. By the last quarter of 2007, cereal and oil prices were rising in lockstep. But cereal prices held steady at their bottom (only a slight increase up to June 2009) whereas crude oil prices continued to plunge. Why? A typical explanation is:

> International food prices did not fall all of the way down to where they had been before the price spike began because a number of countries, including China, were buying heavily to rebuild their food stocks just in case international prices began rising again. Paarlberg (2010, 23)

Others like the prestigious International Food Policy Research Institute in Washington, DC chimed in with the following explanation:

> strong demand from energy exporters keeps rice prices 25–30% above 2007 levels Headey and Fan (2010, 45).

The correct explanation is that biofuel mandate price premiums kicked in (although stock rebuilding does increase mandate premiums, but so do all other commodity market developments—how to apportion?). Food grain and oilseed prices "floated up and away" from the *lowest* possible prices (those occurring when *locked* onto biofuel, fuel, and, therefore, crude oil prices). The mandate began driving the price of corn, which in turn affected all cereal and oilseed prices.[3] Ethanol (and, therefore, corn and soybean prices) increased above that dictated when ethanol prices were directly linked to crude oil (through gasoline and diesel) prices plus the tax credit in the price run-up and collapse of 2008. The mandate premiums over the tax credit were slightly negative in November and December 2008 but rocketed to $1 per bushel in January 2009 (see Figure 3.2 in Chapter 3). Although the mandate premium went to zero for several months in mid-2010, it went close to $2.00 per bushel by November of that same year (and close to $2.50 per bushel in the last half of 2011—see Figure 3.2 of Chapter 3).

Before turning to the specifics of corn and wheat prices, let us first evaluate cereal and crude oil prices in Figure 4.4 for the time period December 2008 to June 2010. December 2008 marked the nadir in cereal and crude oil prices right after the 2008 price peaks. Note that crude oil prices went down farther than cereals from the peak. However, crude oil prices went up from $41 per barrel to $75 per barrel from December 2008 to June 2010 (basically straight up). But cereal prices held steady from December 2008 to June 2009 and then slowly fell although crude oil prices were rising the entire time period from December 2008 to June 2010. Why was there a negative relationship between crude oil and cereal prices? The answer is mandate premiums over the tax credit were being built. Our theory of mandates in Chapter 2 makes clear that, when the mandate is binding, one expects a weak negative relationship between crop and energy (and hence crude oil) prices, ceteris paribus. This means biofuel prices have a bigger impact on cereal prices as biofuel policy price premiums over the free-market gasoline and diesel prices increase; in other words, when one expects a negative relationship between crop and crude oil prices! This will be a key insight when evaluating the econometric literatures in the Chapters 5–8 in this book.

But, after June 2010, cereal prices reversed their decline and marched straight up with no hesitation. This climb came alongside of crude oil price increases, peaking at $123 per barrel. However, this time, cereal and crude oil prices were strongly positively related. There is a strong positive relationship between cereal and crude oil prices for the June 2010 to April 2011 time period (see Figure 4.4). Over this period, mandate premiums were zero from February to August of 2010, but went up to peak of $2.00 per bushel in November 2010. The premium gradually fell to zero in April 2011, the last month of the time period we are now analyzing. Thus, for part of the time, one would expect a strong positive relationship between cereals and crude oil prices.

From April 2011 to June 2012, mandate premiums were high. However, the premiums began falling to low levels from February 2012 to April 2012. The US drought hit in June 2012, making corn scarcer. This, again, led to large mandate premiums that lasted until September 2013. Although high corn prices persisted from September 2013 to March 2014 as shown in Figure 4.2, we showed in our discussion of Figures 2.3 and 2.4 that corn price fell to levels corresponding to the "lowest when locked" onto crude oil prices over this period because of ethanol blending constraints (due to bad weather and rail car shortages) and a high relative price of dried distillers grains with solubles [DDGS; itself perhaps due to ethanol-blending constraints as few ethanol plants were coming on line because the Environmental Protection Agency (EPA) in November 2013 reduced the mandate by 1.4 billion gallons for 2014].

In July 2010, crude oil prices rose substantially and so too did cereal prices. Going back to the corn–wheat price movements depicted in Figure 4.2, beginning in December 2008 when both prices reached their lows right after the precipitous fall from the peak (with wheat falling somewhat more than corn), both prices trended up until June 2010 and then trended down until June 2010 (with wheat going down more and testing their lows of January 2005—corn prices stayed at these levels after they had reacted to the ethanol capacity doubling in 2006). However, in July 2010, all of a sudden, both corn and wheat prices marched up in tandem because oil prices rose (see Figure 4.4). Yet, crude oil prices were rising since December 2008. Why did corn (and soybeans) wait until now to make a move? The reason is that, as oil prices rose, mandate premiums were being eroded—the corn price implied by the oil price was rising whereas the price required by the mandate was declining until the premium disappeared. So, corn and crude oil prices moved together. This time, wheat did not race ahead of corn or go higher than corn; in fact, wheat prices did not rise as much as corn. However, corn went higher than ever although crude oil prices had not risen as much as in 2008. Corn prices peaked in April 2011 at a higher level than in 2008 (because of mandate premiums this time around). But, during the period after corn reached a new peak in April 2011, wheat and corn prices continued to follow each other closely. These prices continued to follow one another even after the drought in 2012 when corn hit an even higher peak. In late 2013, relative prices returned to historical levels.

As depicted in Figure 4.4, after June 2010, oil prices began their march right back up again, but this time, corn prices led the way with premiums due to the mandate (see Figure 3.2). Corn (and, therefore, all other crops that are direct substitutes in demand and compete for the same land in production) led oil prices in 2010 and beyond (as shown in Figure 4.4 with the cereal price index). Corn prices had no choice but to follow suit because ethanol prices dictated it (or vice versa in 2010 onwards as the corn prices have more influence on ethanol prices when the mandate is binding).[4] Wheat prices touched January 2005 lows in mid-2010, but quickly reversed and followed corn prices upwards as oil prices began to rise and coarse grain and oilseed prices were now at a premium to oil (versus a discount in 2008) and, therefore, rose even faster compared to 2008 (see Figure 4.2). Thus, corn and oilseed prices peaked in April 2011 at a higher price than in 2008 with lower oil prices. Wheat prices followed as did rice prices (see Figure 4.3), after some hesitation during the October 2010 to June 2011 time period. Although the very high ethanol price premiums due to the mandates beginning in October 2008 evaporated several times, the mandate premium is at a record high in 2012 because of the expired US blender's tax credit and drought, thereby generating a large gap between the cereals price index and oil price (Figure 4.4).[5]

Meanwhile, it is clear that changes in wheat and rice prices had no direct effect on coarse-grain or oilseed prices in 2008. But, were changes in wheat and rice prices totally independent of coarse grain and oilseed (and, therefore, crude oil) price movements? We argue that the two must have been related. Rice prices did not change significantly until India's rice ban took place in October 2007 (Figure 4.3). As the rice price was, by far, the last of the food grains to react to overall market conditions, it nevertheless surpassed the corn price at $5.30 per bushel in March, 2008. Meanwhile, the wheat price had exceeded the corn price earlier, in July 2007, after corn had almost doubled to approximately $4 per bushel in February 2007. Subsequently, both descended from their 2008 peaks before the corn prices adjusted downward. For a time, wheat and rice prices descended even further.

However, after mid-2010, corn (and, thus, soybean and other oilseed) prices re-emerged as the price leader, and resumed following crude oil prices. Not surprisingly, wheat and rice followed along: both compete for land with coarse grains and oilseeds in various parts of the world, whereas wheat can be fed to livestock and is substitutable for rice in human consumption (e.g., in India).

Wheat and corn prices were well on their way up toward their peak by the time India imposed a ban on rice exports, causing rice prices to suddenly spike and actually overtake both wheat and corn prices. A better than expected wheat crop resulted in wheat prices declining until March whereas corn and oilseeds followed oil to their peaks in June of 2008.

The key questions we must answer in order to determine the full impact of biofuel policy are (1) how much of the developing country policy response that hit the rice market especially hard (with panic, speculation, and other non-fundamental market forces) was due to coarse grains and oilseeds prices being tied directly to oil prices—a price link that became especially tight after September 2007—and (2) how much of the wheat and rice price increases were independent of coarse grains and oilseed prices and, therefore, of biofuels policy influence? Wright (2003) argues it was low rice stocks and uncertain rice supply that caused developing country policy responses and the resulting high rice prices, independent of biofuels and, therefore, of corn prices. But our hypothesis, advanced in this book, is that biofuels policies were a key instigator of the sharp price rises in 2006 and onwards.

The 2007–2008 Price Boom and Beyond: An Autopsy

Biofuel policies and corn markets started the ball rolling and spilled over into the wheat market (wheat could not and, therefore, did not spill over into the coarse grain/oilseed prices; the latter prices were locked onto crude oil prices—there was no choice; this is the nature of crop–biofuel price links in this new era), which, in turn, spilled over into the political decision making of Asian governments, causing policy responses and speculation, including hoarding by individual consumers, which allegedly caused rice prices to maintain its historical values with wheat and corn (see Boxes 4.1 and 4.2).[6]

It is clear in examining the corpse of the 2008 price boom that the fingerprints of biofuel policies were all over it In no way could the crop–biofuel price link have been established without prior biofuel policies. High crude oil prices in 2008 without any policy would have incentivized ethanol production but not in time to establish the critical price links by 2008. That may have occurred in the following years but not in 2008. Negative levels of "water" in the ethanol price premium do not indicate there would have been free-market levels of ethanol production in 2008.[7] Rather, it would take sustained expected profits to have established the original investments in capacity necessary to establish the price link. The entire price corn/soybean price increase happened due to the crop–biofuel price link, itself established because of past biofuel policies. In fact, were it not for biofuel production constraints, we showed in Chapter 2 that, given the level of crude oil prices, corn and soybean prices would otherwise have been higher.

So, the question becomes how much of the wheat and rice price increase is due to the law of one relative international price of grains and oilseeds? If 59 percent of world cropland is locked onto crude oil prices (with no choice), could wheat and rice prices have stayed put? This asymmetry has been overlooked. Our approach is to determine how much of the price increase was due to the corn (soybean) price link to crude oil prices. The literature starts with quantity.[8] We can isolate how much quantity shifts would matter. If prices were locked all of the way up and all of the way down, then we know quantity has no effect. Ignoring the law of one price and believing that, if it was not for the perfect storm factors, wheat and rice prices would have stood by and watched coarse grains/oilseed prices soar is incredible.

Clearly, there was confusion, with something happening that was unprecedented. One way of looking at this is to ask what would have happened if some authoritative and credible body announced in advance that crude oil prices were going to $147 per barrel and that feed grains and oilseeds would be locked onto these crude oil prices (with high multipliers) but at a discount because of ethanol production capacity constraints. Let us assume the market believed it. How would wheat and rice prices have reacted? Probably exactly like they did when corn and soybeans prices reached an even higher peak in April 2011 (and again an even higher peak in July 2012): obeying the law of one price but keeping their distance, given the mandate premiums for corn and soybeans (see Figures 4.2, 4.3, and 4.4). These were somewhat different responses than the one coinciding with the first peak in corn prices in 2008. In fact, in hindsight, it seems all of the responses are reasonable or what would be expected, given the circumstances.

Nevertheless, blaming low stocks, a speculative bubble, flooding and droughts in different parts of the world, exchange rates, easy money and low interest rates, the "financialization" of grain/oilseed markets, the searing income growth in highly populated Asia eating more meat, sharp increases in crop input costs due to high crude oil prices, bans on genetically modified crops, global warming, and the umpteen other perfect storm factors for the increase in wheat and rice prices in 2008 is equally incredible. Ignoring the key price links between corn and ethanol (and between soybean oil and biodiesel) with high crop–biofuel multipliers and, therefore, not taking into account the fact that the entire price increase for the coarse grain/oilseed price would have occurred regardless of any of the perfect storm factors is the most incredible omissions of all by the literature to date; in fact, it is pure ignorance and borders on fantasy. To this literature we now turn.

Box 4.1 Wheat's conversation with corn

February 2007: "Corn, why did you go up so much? For heaven's sakes, nothin' was happinin'!" And yet Illinois farm prices rose 88 percent, white corn up 107 percent. "Oh yes there was," replied corn, "ethanol capacity doubled last year and I was getting hooked onto crude oil prices for the first time in my life (through ethanol and gasoline prices of course) and with a big multiplier to boot. I was getting ready and because we compete with you for land and are substitutable in consumption (not to mention you are being used for ethanol production in Canada and Europe), you had better join the ride too!" "Nonsense," replied wheat: "nothin' happening. In fact, I am going to drift down for a while, just to make a point."

October 2007: "Hey wheat, why you going up now? Trying to make up for lost time and keep up with me, eh! Catching on that we are in a whole new ballgame, uh!" "Nonsense," says wheat, "we got some Midwest flooding and a drought in Australia." "Woopdy doo," scorns corn. A little while later, corn exclaims: "Hey, you have caught up to me! I am at $3.84 per bushel but the zero profit condition in that de Gorter and Just equation says I should be at $5.14 a bushel. Drat—hope that does not mean you pass me by. But anyway, crude oil prices are going straight up and so am I. I ain't got no choice. Those ethanol boys have me locked in. How about you? Are you just going to hang around here?" "No, in fact, I am going to say cheerio, my darling, and head on up," chimed wheat. "I got exchange rates, speculation, the new financialization of commodity markets, higher input costs, I got some unique 'market structure' features, heck, I am in a Perfect Storm! Everything is swirling around me." "Perfect storm!", exclaimed corn. "Nothing matters except the crude oil price. Me and soybbies are locked, baby. Low inventories, Chinese eating more meat, global warming, whatever, nothin' matters, honey. We is locked onto crude oil prices." "Well, you got your own perfect storm: a strong but steady gale wind in the North Atlantic (with some cross currents holding you back, namely ethanol production constraints)," replied wheat. "It is perfect nonetheless, as simple as it is. But my storm is in the south Pacific and it is unique: every factor in the world matters for me (except you!), and it is vicious—tropical storm like."

And so wheat marched up, past corn although corn and wheat should have gone up hand in hand if there were no ethanol production-capacity constraints, according to the de Gorter and Just formula. However, wheat had overreacted and its price peaked in March 2008 and so headed south to get back in "equilibrium" with corn, and crossed corn marching north at a steep angle, locked onto crude oil prices. "Hey, where are you goin' corn?", exclaimed wheat. "Where am I goin'!", exclaimed corn. "I am followin' crude oil up—how many times do I have to explain that to you!?" "My bad weather eased, exchange rates are going the other way around, demand is shifting in because of the recession, global warming is a myth, all the perfect storm factors are unwinding," replied wheat. "I am goin' down now." And so wheat departed ways with corn (and soybeans) following crude oil closely all the way to the top (although wheat prices formed the same mini "mountain cap" as corn prices in the three months of May–July 2008).

July 2008 to 2014: Corn followed crude oil prices all of the way down; halfway down, wheat and corn joined hands and dropped to their lows in the same month as crude oil prices, except wheat unlocked her grasp with corn in August 2009 and settled at a relatively lower level. They went up for a while and then drifted lower. But there was perfect co-movement—harmony like in the past. In the first five months of 2010, wheat drifted lower at a faster pace. This made corn anxious. "Hey, where you goin', exclaimed corn. "I goin' right back to where I began in January 2005," replied wheat. "This is the worst economic depression since the Great Depression. There is nothing justifying my high valuations." "But there has to be some flooding somewhere," exclaimed corn. "Some regional droughts. There always is. Can't you come up on that alone? I have no choice here. I have to stay up here, mandate premiums and all that. I cannot go down, crude oil prices have been climbing for months on end and as soon as mandate premiums dissipate, I go straight up with crude oil prices. I am in a straitjacket. You will be sorry."

And sure enough, in July of 2010, up went corn (2.4-fold in the following 12 months) but this time wheat did too (2.2-fold from its recent low). Corn asked wheat why and wheat said she had learned her lesson—never again would she ignore corn and then have to go through the hassle of making up, even overreacting. So wheat promised that anywhere corn goes, she would follow but she did not tell corn she would keep her distance when the mandate was binding—those mandate premiums that corn prices had to absorb were a little too heady for her. And wheat promised one more thing: if corn got to a point where it was *lowest* when *locked*, she would meet with corn head on. And lo and behold, corn prices got to its *lowest* when *locked* onto crude oil prices in September 2013 and from that time on, corn and wheat were back again, hand in hand, like the good old days. Until the next storm!

Box 4.2 Wheat's interview with rice

Wheat: Hello rice. It is well into 2014 and I just wanted to have a chat on what happened a long time ago in 2008.

Rice: Nice to be with you wheat. Ancient history was my favorite subject in school. Glad to talk.

Wheat: Rice, your prices rose 15 percent in the first quarter of 2007, having drifted sideways since January 2003. Why?

Rice: Ah, just a sympathy move with you (up 33 percent since January 1, 2005) and then corn had their burst in October 2006. Nothin' else. Not sure what was goin' on with corn but you had some bad weather so if other markets move, we do too. Your stocks were somewhat depleted; our governments had run our stocks a little low. Besides, co-movement is a statistical reality: exchange rates, incomes, etc., and, besides, the law of one relative international price for grains and oilseeds is alive and well so if something happens in one market and we are not sure what it is, traders arbitrage what would otherwise normally develop into excess profits if left unattended to. Some people call it speculation; others call it anticipation. This is how markets work, watching relative prices and arbitrage.

But then come June 2007, your prices jumped and corn yelled at us: "Hey, better come up with us; something is happening in the biofuel world and if crude oil prices maintain their rise, we is all goin' up!" I have to admit we just laughed at corn. I remember saying "Biofuels are just a fad." We did not think it would affect corn and soybean prices very much, let alone cause a ripple in our market. After all, biofuels accounted for less than one percent of total cropland back then.

But you and corn were right. That crude oil price jump whacked feed grains and oilseeds right in the pants and it was all up from there. You (wheat) got on the bandwagon pretty early and headed north based on the hypothesis you were in your own perfect storm. I kind of doubt you were. We (rice) waited until the last minute possible to go up. We were surprised: our per capita demand continued its slight long term decline. No production shocks. Stock levels a little low relative to what we were used to in the past (which were probably a little too high) but basically fine. Nothing was happening. All quiet on the eastern front. But once your prices were tripling, we is going up.

Wheat: Some economists just do not believe it was speculation that caused your prices to go up.

Rice: Well, we went up threefold in a flash and everybody and every market participant including governments was "speculating" on what was happening or trying to predict what was going to happen next. Hedge funds don't trade commodity index funds to hedge; they speculate. State trading enterprises try to predict; some would call that speculating (certainly not a science with certainty what the future will be!) Ditto national governments; they react in all kinds of ways; have to keep face. Peasants hoarded rice. They were confused, uncertain like everybody else. Some call it speculating; others call it prudence (same thing could be said for what everybody else that participated in the rice market at that critical time). The minute the US agreed to allow Japan to release rice stocks, the market plummeted. And Japan never did release any stocks. Some call it speculation; others call it policy response. Whatever. Not normal. No standard economic fundamentals at work. Need a new theory of price determination.

Wheat: Are you saying biofuel policies may have had a bigger impact on rice prices than for corn in 2008?

Rice: Yes, you can interpret it as such, putting everything together.

Wheat: But when corn hit new highs in 2011, your reaction in terms of price change was more moderate. Why?

Rice: Ah, we learned. We understood what happened after the fact. Peasants did not hoard rice the next time around. They knew it was not a perfect storm; basically biofuels and a feed grain/oilseed thing mainly. We co-moved, do not get me wrong, but kept our distance and let economics and the law of one relative price do the work.

Wheat: Mmm. Peasants knew it was not a perfect storm. Are you saying peasants are smarter than so many economists who say otherwise?

Rice: No, not saying that but you have to admit that a couple of billion people who spend more than half their income on basic foods have more of an incentive than economists in academia and elsewhere to understand what is going on. Their life depends on it.

Wheat: Very interesting. Thanks so much.

Rice: My pleasure.

CHAPTER 5

A CRITIQUE OF THE LITERATURE ANALYZING BIOFUEL POLICY AND THE 2008 FOOD COMMODITY PRICE BOOM

Economists know all about prices and nothing of value.

Anonymous, graffiti in a second-floor bathroom, Giannini Hall, University of California, Berkeley, Fall 1979.

Most academics analyze issues through the lens of their own expertise. Therefore, it should be no surprise when a recently minted Nobel Prize-winning behavioral economist wrote in the New York Times that:

> Commodities followed the euphoria cycle that we had along with housing. Shiller (2008)

Behavioral economists routinely simulate stock market bubbles and have fixated on such bubble phenomena. Figure 5.1 does not lend a lot of support for his assertion as the supposed bust was followed by an immediate return to the run-up. In the run-up of 2008, the crude oil market may have experienced euphoria, having direct implications for the grain/oilseed prices given their price links. Nonetheless, euphoria and bubbles represents a theoretical framework that has limited use to explain the overall issues we are examining in this book.

Wright (2011, 2014), a foremost expert on the competitive theory of storage, who has written many books and papers over 30 years on this important aspect of grain/oilseed markets, emphasizes stockholding behavior and how low stocks with supply/demand shocks generate "bubble"-like price behavior. It turns out low stocks do matter (just as supply/demand shifts ignoring stocks), but only up to the point when the mandate

Figure 5.1 Housing versus corn prices.

Source: S&P/Case-Shiller 20-City Composite Home Price Index, USDA Feed Grain Yearbook Table 12.

premiums dissipates (which we showed in the previous chapter is only 20 percent of the food commodity price increase). Thus, we will reserve a full discussion of this theory of storage to Chapter 8 where we will incorporate it into a more complete model of biofuel prices. However, by itself, we would regard this theory to be a dubious explanation of market movements since 2006.

Many economists hypothesize that macroeconomics has been the key driving factor. Overshooting exchange rates, easy money (low interest rates), portfolio diversification (need to put a little corn with your copper in your retirement fund), and many other so-called realities of the day dominate the thinking of these economists. Frankel (2008) is a case in point declaring "Commodity prices are back, with a vengeance." He and those like him dwell mostly on monetary policy and interest rates as the primary driver of the commodity price boom. Figure 5.2 plots the US commodity-weighted exchange rate against the corn price with a correlation coefficient of −0.83 (although the exchange rate changed a whopping 10 percent from January 2006 to January 2010). The diagram reflects that the changes in commodity prices were not simply nominal responses to changes in the money supply, among others. Rather, this was a real increase in the world value of agricultural commodities.

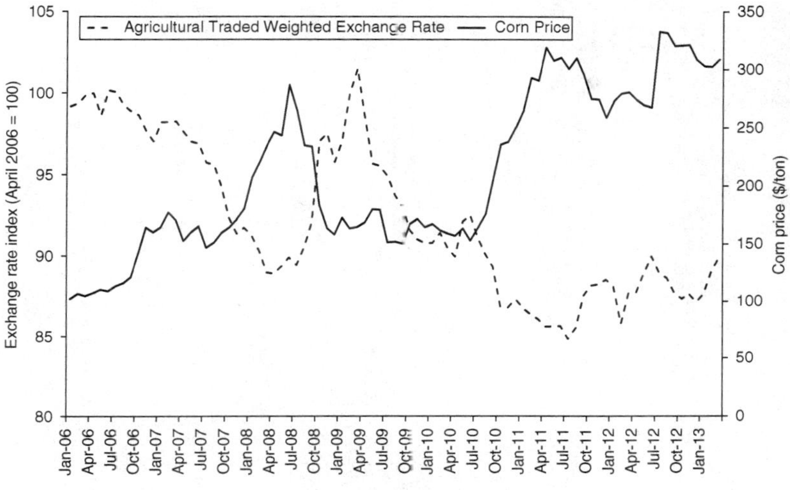

Figure 5.2 Exchange rates versus corn price.

Source: World Bank (2014), USDA Agricultural Exchange Rate Data Set.

Economists at the International Monetary Fund (IMF) are another example of relying on the familiar. One of the more eye-lifting papers is that of Cevika and Sedik (2014):

> Analyzing two very distinct goods—crude oil and fine wine—helps to identify common determinants of commodity prices...global aggregate demand is the key macroeconomic determinant of the fluctuations in both crude oil and fine wine prices.

So, people just wanted more of everything; but this is more of an observation rather than an explanation. Indeed, fine wine, grapes, beer, barley, whiskey, rye, bourbon, corn, sake, rice, vodka, wheat, ____, soybeans—it always comes down to the same thing, does it not? Some economists emphasize "China's soy tsunami" where one of the most oft-quoted statistics is the acres of soybeans imported by China (versus the acres devoted to biodiesel, which is always less). The correlation coefficient between soybean prices and imports of soybeans to China is 0.88 (see Figure 5.3). This tsunami combined with the correlation coefficient for exchange rates and the cereal price index from Figure 5.2 was said to be more than enough to explain 100 percent of the food commodity price increase. A recent sign in a downtown Toronto church board "do not let facts interfere with the truth."

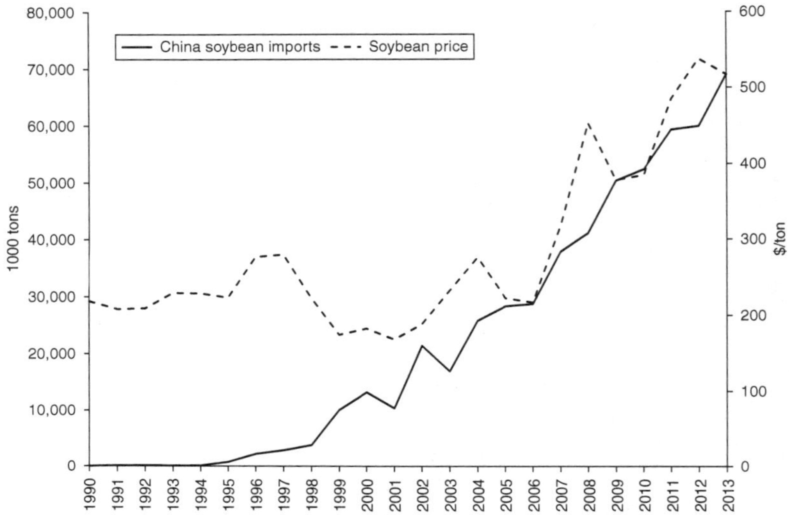

Figure 5.3 China soybean imports and soybean prices.
Source: Index Mundi (2014).

Clearly, Figures 5.2 and 5.3 are facts read, but we will show that Chinese eating more meat and macroeconomic phenomenon like exchange rates had little to do with the food commodity price boom.

To add insult to injury, the IMF recently sponsored a conference on commodity prices and published 17 papers in a special issue of the *Journal of International Money and Finance* in 2014. But, in only one paper are biofuels even mentioned. The abstract of Avalos (2014) reads:

> This paper steps out of the literature's typical emphasis on macroeconomic drivers...results show the transmission of oil price...to corn prices has become stronger after 2006 (no changes with respect to soybeans)...also a significant transmission of corn price...to oil and soybean prices.

Getting away from the "typical emphasis on macroeconomic drivers" is one thing we endorse. However, we believe "the transmission of oil price...to corn prices has become stronger after 2006" is perhaps not nuanced enough to satisfy reality. In some time periods, we think this statement is correct (some days, weeks, or months in January 2007 to December 2008 in Figure 4.4 and again from June 2010 to April 2011); in other time periods (as we saw for December 2008 to May 2010), we would expect the opposite in the appropriate sub-time periods. When one accounts for how the price relationships must work, the relationship depends. There could be a positive or negative relationship between corn and crude oil prices. However, Avalos (2014) makes no adjustment; in fact, he makes no mention of a mandate versus a tax credit or that there is more than one regime.[1] Rather, he—like many before him—employs a bare bones statistical analysis on monthly data with little more than a hunch and no mention of our literature. And why does he obtain different results for corn versus soybeans? We show they went in tandem pretty much the whole way from September 2006 to now. And what is this about corn transmitting to crude oil prices? Is this an endorsement of the Du and Hayes (2012) results? (See Box 3.1). The approach taken and the results obtained in Avalos (2014) do not instill a lot of confidence.

The overriding themes of the literature are that the 2007–2008 price boom and ensuing volatility was caused by a "perfect storm" of coinciding events. This "perfect storm" analysis was based entirely on the notion that conventional economic models (ignoring the new economics of biofuel policies as developed in this book) are more than sufficient to explain all of the price swings. They argue that there is no reason to go "far afield" as we have (and will continue to go in later chapters). These fundamentalists also heap scorn on economists who believe this time is different—that the 2008 price boom cannot be explained by "fundamentals" or conventional models [see, e.g., Wright's (2014) critique of what we consider excellent analyses in various papers by Timmer, Dawe, and Slayton who argue conventional models cannot explain the price boom]. There have been some qualitative studies that give proper weight to biofuels [e.g., Runge and Senauer (2007); Mitchell (2008) as we discussed in the preface of this book; Collins (2008) who used part of our formula; Timmer (2010) and Rausser and de Gorter (2012)]. But as in all other studies to follow, each study examines different scenarios, baselines, amounts of ethanol consumed, and chooses different projection years, not to mention other variables (Condon et al. 2013; Zhang et al. 2013). Thus, they are very difficult to compare objectively to each other or to the results we present in this book.

How to Get a Handle on the Literature?

Surveying the literature and making it understandable is no easy task, as witnessed by a recent FAO study on biofuels and food security (HLPE 2013, 7–8):

> An international, intergovernmental and multi stakeholder discussion on biofuels and food security could well start based on a jungle of more than 1,000 documents and reports. That would not be very useful…leav[ing] to each

> policy-maker the difficult task of selecting suitable and relevant documents, making his/her own synthesis, necessarily partial, reading the arguments of all the other parties, trying to understand them, and trying to make his/her point in the debate by articulating his/her view to all the others.

Wow! No kidding. The FAO report seems to be in despair over the task of having to systemize a literature that seems to be unsystematic in itself—all those strange animals in the jungle of papers pretending to understand the role of biofuels in affecting food commodity prices.[2] There are at least 1,000 papers, studies, and book chapters written on the economics of biofuel policies and their impact on food commodity markets alone.[3] However, this chapter in our book, as a courtesy, will isolate the "suitable and relevant documents" in this jungle and impartially synthesize these papers with a full understanding of all the arguments. We will also quite quickly dispose of the other "unsuitable and irrelevant documents" (which are in the majority, so it will make our task easier than otherwise—reviewing all of these 1,000+ papers would no doubt double the length of this book). In this way, we will be able to articulate the central role biofuels had on the level and volatility of food commodity prices, with a guarantee that the reader will make a giant leap toward understanding this important issue. Where possible, we will link the literature's result to the model we have already expounded (although, in most cases, it will be painfully clear and unnecessary, or at least require minimal explanation). Sometimes, our model will obtain diametrically opposite results from the literature. At other times, our model is observationally equivalent to the literature. Sometimes the results in the literature are inconsistent. At others, our model can predict the price levels but the volatility of prices may be explained by a combination of factors that makes us unable to distinguish which force predominated.

Donald Rumsfeld has been quoted for saying:

> There are known knowns and known unknowns. But there are also unknown unknowns—ones we don't know we don't know.

The literature is really an unknown known: it is severely lacking in that it tends toward a disparate collection of extraordinary explanations for unique phenomena. We attempt to replace this failed literature with a coherent explanation of the entire array of phenomena.

The unknown unknowns are probably the biggest problem economists studying the economics of biofuels have to overcome in regard to understanding the literature. In this jungle of papers and reports in analyzing the literature on what role biofuels had on food commodity markets, economists continue to use old frameworks of analysis for this new issue of biofuels.

The literature emphasizes how the quantity of biofuels impacts food commodity prices. They ignore the overriding effect of the new price links, established because of biofuel policies. These biofuel policies may have a big impact regardless of how much biofuels are being produced. Moreover, by ignoring these links, many economists when examining 2008, have no idea how little supply and demand shocks mattered.

The market effects of biofuel policies have become one of the most important and hotly debated policy issues of our time. The economics profession responded with a myriad of literatures in trying to explain the impact of biofuel policies, all contradicting each other in methodology, and more so in outcomes. This reflects the chaotic quagmire in economic thought on the analysis of agricultural commodity markets.

Fueling the debate are four different and separate literatures (apart from the one developed in this book)[4] analyzing the impact of biofuel policies on grain and oilseed markets. They include: (1) modeling the effects of biofuel policies as a simple shift in crop demand in a Marshallian supply/demand model; (2) extending the Marshallian framework to include the demand for storage and the resulting implications on both the shape and placement of the total (consumption plus stock) demand curve; (3) econometric time series models that regress quarterly or annual data with the commodity price as the dependent variable while the right-hand side variables include macroeconomic variables such as gross domestic product (GDP) and exchange rates, commodity supply/demand shifts, and the like; and (4) a literature using high-frequency data (daily, weekly, and monthly), frantically seeking for links among crop (corn, sugarcane, and oilseeds), biofuels (biodiesel and ethanol), energy (gasoline and diesel), and crude oil prices using state-of-the-art time series econometrics technique—but ignoring the modeling concerns of all three of the other literatures. Finally, our own literature that we developed in this book relies on the use of standard microeconomic modeling of profit maximization in competitive markets to determine how the market will respond to both policy shifts and external shocks. As the reader will understand by the end of this book, this last literature is the most important among them all.[5]

Each of these literatures is at odds with one another. Approaches (1) and (3) tend to find small impacts of biofuel policy, whereas approach (2) finds that storage demand leads to a steep kink in the demand curve such that the higher prices resulting from biofuel policies have also changed the elasticity of demand substantially. Thus, those following approaches (1) and (3) are much more likely to blame rises in grain prices beginning in 2007 on a host of amazingly coincidental factors with biofuel policies playing a minor role. Approach (2) suggests that biofuel policies may be the primary culprit behind rising grain prices as grains diverted to biofuels escalated in volume.

Finally, approach (4) that analyzes price links only has three sub-literatures: analyzing the levels of prices (relationships among trends) and co-movement of prices; testing for and finding dates for structural change; and transmission of volatility from one price to the other. The first two of these sub-literatures are evaluated in Chapter 7; the topic of volatility is delayed until chapter 8. The results of literature (4) are all over the map (e.g., either no or strong relationships among crop, biofuel, and energy prices) and the reasons for this are various and will be explained in Chapters 7 and 8.

Of these four literatures we have identified (apart from our own and those who have adopted our approach as presented in this book), the first two take "quantity" as the focus of attention and rely on basic economics of commodity price behavior developed over the past 200 years or so. The approach that models biofuels policies as an exogenous shift out in the demand curve can rightfully be thought of as a sort of *manna from heaven approach*. This approach ignores any relationship between the ethanol and corn markets that would occur when blenders seek to maximize their own profits, but rather see blenders as a group of consumers created magically by policy that has no motive other than to buy a particular amount of corn, thus moving the demand curve. By failing to model the vital price links developed in this book so far, they create a false counterfactual of no biofuels with which to compare the current situation. The competitive storage approach also is based on quantity, underscoring the importance of the stocks to use ratio in determining how quantities respond to prices.

The last two literatures take "prices" and rely on state-of-the-art econometric techniques developed in recent decades like testing for co-integration and using Granger causality tests. These models also borrow from the finance literature to assess volatility and varying structural break analysis. Treating each price series as simple draws from distributions that are possibly correlated, this literature then supposes whether there is a simple correlation that the prices of the two markets in question are linked, and if there is no correlation they are not linked. This approach ignores the fact that such price links may only be activated under certain circumstances (e.g., when prices are high enough to justify ethanol production for fuel use, and where the mandate is not binding). Such an exercise has no power to demonstrate that there is no relationship between the markets. In fact, by ignoring the periods in which there should be no relationship, this approach produces results heavily biased toward finding no relationship between corn and fuel prices. Moreover, these authors then take this apparent lack of a relationship as evidence of a negligible effect of biofuel policy. Or vice versa—if there is a strong relationship, say between corn and crude oil prices, then biofuel policies have their largest impact. Nothing could be further from the truth. If the lack of relationship in a particular period is due to ethanol prices being too low to justify production, then such a conclusion is justified. However, if the lack of a relationship is because the mandate is binding (as it has been the case in much of the last five years), then the policy has the largest possible impact (on corn prices).

This last literature, which is desperately seeking a relationship between prices to determine if biofuels are a factor in the food grain/oilseed price booms, operates much more like a fishing expedition. These authors randomly choose which prices to consider: corn to ethanol, corn to gasoline, or corn to crude oil? Further, they make no distinction that a corn to ethanol relationship is entirely different than corn to gasoline or crude oil price relationship. Not one paper of the umpteen papers even mentions the possibility that tax credits versus mandates may impact price relationships (which we have shown so far determines the sign between a crop and an energy price!). Moreover, many of these papers try to draw conclusions about which commodity is responsible for the price run up and which were simply responding based on relative timing of responses. Unfortunately, this approach ignores the potential for forward-looking investors. If corn futures prices rise before oil prices, it may or may not reflect an influence of corn on oil. One alternative explanation may be that both are responding to the same market information at the same time, but that idiosyncratic error leads the corn market to respond first. We delay discussion of this particular literature until Chapter 7.

Of course, many papers employ a combination of the several approaches identified in this chapter. For example, some studies emphasize the market dynamics of stockholding and the role of a shift in demand for corn (and soybeans/rapeseeds) due to biofuels, while using methodologies in each of the literatures described (Timmer 2008; Abbott et al. 2009; Hochman et al. 2011; Wright 2011; Byrd 2012; Carter et al. 2013; Janzen et al. 2014).[6]

The "Perfect Storm" Explanation

The most popular theory explaining the price surges amongst economists is the near-"perfect storm." As Taleb (2012, 221) puts it:

> No, we don't put theories into practice. We create theories out of practice ... it is easy to infer from it ... that the confusion is generalized.

The "perfect storm" theory relies on the interaction of a complex set of coincidental factors. This approach maintains that some cosmic roll of the dice led to a confluence of macroeconomic forces (economic growth, weak dollar, fiscal expansion, low cost of capital, and speculation with the new "financialization" of food commodities), sector-specific factors (higher input costs due to energy prices, bad weather, crop disease, global warming causing droughts, changing Asian diets, low stocks, and, oh yes, even biofuels!), and policy-related factors (like developing country policy responses, preexisting farm policies, and low public R&D in agriculture) to suddenly and inexplicably begin to impact commodity prices at once.[7]

This literature necessarily argues many things happened coincidently, with a demonstrable overreliance on the word "complex." There are various sub-genres of this near-"perfect storm" literature. There are both qualitative analyses, modeling biofuels in partial and general equilibrium models as 'manna from heaven and price dependent econometric time series analyses using macroeconomic variables and stocks to use ratios as regressors.

Table 5.1 The perfect storm explanation for the food grain/oilseed price boom

Economic model	*Economic variables*
{1} Behavioral economics	Speculation, the new financialization of commodity markets...
{2} Macroeconomics	Excess liquidity, GDP growth, fiscal expansion, portfolio diversification into commodities...
{3} Exchange rate	US dollar depreciation
{4} Shift in demand curves	Demand shocks due to surging incomes in developing world, changes in diets...
{5} Shift in supply curves	Supply shocks due to increased input costs due to higher energy prices, bad weather, disease, climate change...
{6} Declining agricultural productivity	Public R&D expenditures for agricultural crop production declining.
{7} Domestic and trade commodity policy	
(a) Policy responses to rising crop prices	Export bans and export taxes, lower import barriers, domestic consumer price controls...
(b) Preexisting farm and trade policies	Existence exacerbates the effects of policy responses and of biofuel policies.
{8} The Theory of Crop Storage	**Non-linear relationship between crop price and stocks-to-use-ratio.**
{9} Biofuels	
(a) "Manna from heaven" analysis	Shift inelastic crop demand right (quantity of biofuels arrives from nowhere).
(b) **"The Theory of Biofuel Policy"**	**Careful analysis of what biofuel policy is binding and the specific relations between the energy, biofuel, and feedstock prices domestically and internationally...**

Source: Authors.

Table 5.1 categorizes the different underlying economic models for the "perfect storm" explanation and the second column provides the corresponding economic variables that are most commonly cited in the literature. Table 5.1 is self-explanatory. There is a far-ranging set of economic models, from behavioral economics to "the theory of biofuel policy" {model [9(b)]} that we espouse in this book. We also highlight model {8} (the theory of crop storage) because this model has credence with the unexpected permanent biofuel demand shock depleting inventories, causing prices to rise even further because of the nonlinear relationship between prices and inventory levels. We shall survey this literature in Chapter 8.

Qualitative Analyses

Some papers in this genre argue that it is impossible to determine the role of biofuel policies in the price run up and, thus, we should not attempt even apportioning the relative values of such factors. The causes are so "complex" in this "perfect storm" setting that all attempts are doomed from the outset. The following assessments are, perhaps, most revealing about this literature:

> The factors driving current food price increases are complex. We make no attempt to calculate what percentage of price changes are attributable to the many disparate causes, and, indeed, think it is impossible to do so. Abbott et al. (2008, 5)
>
> Wisely, it does not attempt to apportion the total rise among the different drivers. So, given this complex maze of factors, can we say how much was due to [this, that and the other thing]...The answer is absolutely no...one cannot, with any precision, partition the effects. Abbott et al. (2009, 39) on Timmer (2008)

Abbott et al. (2008, 44) contradict themselves by giving some empirical estimates on the biofuels contribution. This was not presented in the most straightforward manner, and so it is best we use their words:

> Babcock and McPhail estimate that eliminating the ethanol mandate would reduce corn price for 2008 by $0.26 to $5.34, a decrease of 5%. They estimate that eliminating both the subsidy and the mandate would reduce corn prices 14% for 2008. These short-run results are generally consistent with our results presented above.

One may ask why the literature, almost to the last paper, ignores the basic framework of the competitive theory of the firm as it interacts with biofuel markets (as developed in this book)—not even a mention. For example, Abbott et al. (2008; 2009) survey a couple of dozen papers but anything related to this framework of analysis except for Collins (2008) who employs a coefficient like that in the denominator of equation (2.1). Abbott et al. (2009, 42) give the following review of the Collins (2008) paper:

> Using a simple analytical model to impute the role of corn ethanol on corn prices.

This is in contrast to the "complex" and vast set of coincidences they attempt to enumerate but not prioritize. In the least, this is not a ringing endorsement of our approach. We contend that, while it *is* simple, the theory of biofuels is powerful—summarizing the important characteristics of the markets and policies involved, and taming the seemingly complex set of forces in the marketplace since 2006.[8]

Headey and Fan (2010, x) is more of the same:

> Many possible causes have been identified, but their relative importance is uncertain...the result of a complex set of interacting factors rather than any single factor.

Headey and Fan (2010) yet add "despite this complexity, the assessment presented here suggests that some explanations still hold up better than others."

Another study, although it does not try to measure the relative impacts, by Carter et al. (2011, 89) do a good job of covering the basic factors affecting commodity price booms:

> We explore...two largest and most dramatic events since World War II: booms and busts of 1973–1974 and 2007–2008...[there is] no simple explanation for what causes such complex events...these two events are much more similar than different and they provide a context to assess possible future booms and busts.

The paper is far-ranging, including a discussion on the importance of "overshooting," "asset bubbles," and "continued globalization, population growth, urbanization, increased regional specialization of agricultural production and trade, biofuel demand, and climate change." Another insightful passage from Carter et al. (2011, 87):

> We find at the core of these cycles a set of contemporaneous supply and demand surprises that coincided with low inventories and that were magnified by macroeconomic shocks and policy responses...because market participants are likely to respond by increasing inventory holdings and investing in new technologies, we see no reason to expect an increase in the frequency of dramatic commodity booms and busts.

Their conclusion is that, as in the past, restored inventories and technological change will result in business as usual. We will make some arguments throughout our book that returning to business as usual is not that seamless. We delay our discussion of technological change until Chapter 12 where we will argue that the impacts of some technological changes may be different in this new era of biofuel policies.

There are many studies that resort to nearly every possible contributor of the 2007–2008 price boom (e.g., Stoeckel 2008; McCalla 2009; Trostle 2008; and Trostle et al. 2011). Trostle et al. (2011) is illustrative of the literature, where events depicted in Figure 12 in their article that are presumed to have caused the 2010–2011 price boom are reproduced in our Figure 5.4, except that we have added the gasoline price.[9] Trostle et al. (2011) depict a series of supply disruptions (four droughts, one "dryness," one "high temperatures," one freeze, and two cases of rain damage), three policy responses[10] (two by Russia; a ban on wheat exports in the summer of 2010 and the suspension of a wheat import duty in the winter of 2011—so what is it? Does Russia import or export wheat?), and an increase in demand ("importers aggressively buying"). There is an arrow (almost hidden from view) on the top left-hand corner, emphasizing economic growth and exchange rate changes (the US dollar collapsed a whole 6 percent in this time period) with "a rising oil price" squeezed in between. Note that

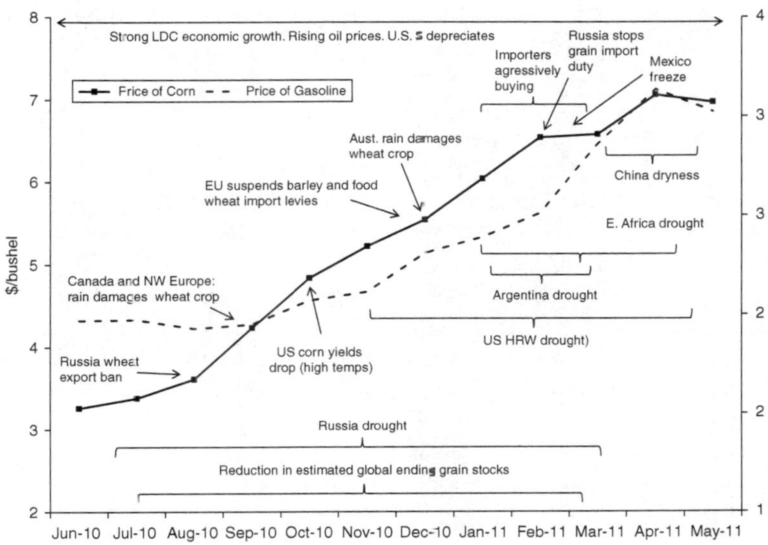

Figure 5.4 Trostle et al. (2011) interpretation of crop price movements 2010–2011

Source: Modified from Trostle et al. (2011).

corn prices rose 2.4-fold in this 12-month time period under investigation (wheat prices rose 2.2-fold).

The abstract of Trostle et al. (2011) explains it all:

> Prices of food commodities increased sharply, surpassing the 2008 peaks that had spread anxiety among policymakers and low income consumers around the world. Most of the long-term trends in agricultural production and consumption that contributed to the 2002–06 price increases and the 2007–08 price spike also contributed to the recent price surge, including global growth in population and per capita incomes, increasing world per capita consumption of animal products, rising energy prices and growing global biofuel production, depreciation of the U.S. dollar, and slower growth in agricultural productivity. The price spikes in both periods also reflect short-term shocks from weather-related production shortfalls, a corresponding decline in world stocks of grains and oilseeds, and changes in trade policies and practices in some countries. Renewed economic growth and demand in low- and middle-income countries...also played a role in recent price increases. While many of the factors that contributed to price increases in 2002–08 and 2010–11 are the same, the timing, sequence, and relative importance of these factors varied.

Notice that there are three time periods identified (2002–2006, 2007–2008, and the 2010–2011) and that the same factors contributed but "the timing, sequence, and relative importance of these factors varied." Moreover, many of these factors have presumably been trending for years if not decades (e.g., population growth, income growth, growth in animal consumption). It is, in the very least, unintuitive that such glacially moving factors could lead to a sudden and rapid price spike.

Trostle et al. (2011) argue that real commodity prices had ended their decline years before September 2006 when, over six months, the price of corn gave a dress rehearsal for the real deal in 2008. The real prices for cereals did not reach their peaks in October 2002 and April 2004 until September 2006 (in other words, real cereal prices were declining going into September 2006—data in Figure 1.3 shows the same for real prices for agriculture as a whole). We recognize that the price surge for grains and oilseeds caught everybody by surprise, as Carter et al. (2011, 92) continue:

> [Due to] the fact that no major single world event marked its beginning, The Economist magazine referred to the 2007–2008 commodity boom as a "silent tsunami.

The reasons why it was silent are (1) the price link created between corn and ethanol was completely unexpected [e.g., unlike China's GDP growing over 9 percent per year since 1978], and the high multipliers were completely unknown; and (2) economists, including *The Economist*, always look for events or "quantities" to tip them off, in typical fashion when using models of Marshallian supply and demand and rational competitive storage. It was a silent tsunami because there were no "fundamentals" in the context of traditional models that ignored the new price links. There is no way one could explain what happened using conventional models that agricultural economists employed. This time it was different.

A series of papers by Dawe, Slayton, and Timmer are particularly insightful and are principally focused on the rice market (Dawe 2009; Timmer 2008, 2010; Slayton 2009; Timmer and Dawe 2010; and Dawe and Slayton 2010). They correctly argue that there was nothing untoward in the rice market in terms of low stocks or changes in fundamentals. The threefold increase in rice prices in the space of just a few months was simply unexplainable using traditional models and "fundamentals." And so they did not pretend to explain this rise either by inventing reasons like a "perfect storm" (or that real prices had stopped declining before September 2006). They simply gave the realities and then made some conjectures and left it at that. They distinguish themselves by pretty much admitting that economists cannot really explain the 2008 price spike using traditional tools of analysis. Therefore, these authors and their analyses are a breath of fresh air. They also provide very thorough analysis of the rice market developments, exhibiting their deep knowledge of the subject matter. We entirely agree with their analysis, which, ironically, extends to all cereals and oilseeds. It is the new fundamentals, the ones we develop in this book, that explain what these authors have correctly identified: there were no underlying quantity-based causes and, unless new fundamental tools are developed, the entire phenomena were largely unexplainable.

Biofuels as Manna from Heaven

The papers discussed so far are qualitative with no formal empirical analysis, and yet we have only chosen a representative few. We now briefly assess the quantitative models that analyze the food grain commodity price increases through the lens of supply and demand shifts, that de Gorter and Drabik (2012a) call *manna from heaven analysis*. This analysis supposes that biofuels just appear from nowhere and shift out the demand curve for a feedstock (e.g., corn) and that is it. No adjustments whatsoever are made to the simple traditional models of analysis to include the insights we have given so far and will expand in the rest of this book. The papers by McPhail and Babcock (2008) and Abbott et al. (2008, 2009) fall in this genre.[11] Other assessments are given by the Organization for Economic Co-operation and Development (2008), Food and Agriculture Organization of the United Nations (2008a), and the Interagency Report to the G20 (2011), just to mention a few from international organizations. Nongovernmental organizations also made assessments. Oxfam (2008) was on the high side with a whopping 30 percent increase in commodity prices due to biofuels! Ecofys' (2013) assessment was much lower. These assessments were essentially mindless applications of the models that had worked up until 2006. By ignoring the fundamental change in the market structure, these analyses were blindsided.

Moschini et al. (2012) survey six of countless Global Trade Analysis Project (GTAP) model simulations of biofuel policies on food commodity prices and six of countless studies by the Center for Agriculture and Rural Development, Iowa State University and the Food and Agricultural Policy Research Institute, University of Missouri (CARD-FAPRI). In all cases, biofuel policies are found to have a small impact on grain and oilseed prices. Zhang et al. (2013) review four studies using partial equilibrium models and five using general equilibrium models. The impacts on corn prices range from a 4.7-percent to a 52.6-percent increase, with an average of approximately 20 percent. A survey by Timilsina and Shrestha (2010) find a wide range of estimates, although the quantitative studies we are assessing here have lower increases than the qualitative analysis.

The EPA, in a study by Condon et al. (2013), review 19 studies and find a range from nil to 72 percent [the latter in Rosegrant et al. (2008) who settled on a 30-percent increase for the price of corn as the best estimate overall]. However, this is apparently at the low end of the weighted average increase in corn prices due to ethanol production. Condon et al. (2013) determined that corn prices increase 2–3 percent for each billion gallons of ethanol production (disregarding our analysis and the point we make in Figure 1.1 that quantity may not always give the correct picture; it is the price links that really matter).

Hochman et al. (2011) survey the literature [e.g., Rajagopal et al. (2009a) found a 15–28 percent corn price impact for 2007–2008]. Chapter 3 of HLPE (2013) provides a survey of the literature although we delineate the literature in a very different way. A wide range of estimates is given again but the normal supply/demand shifts and biofuels as manna from heaven is the general approach used. Moreover, numerous government agencies pondered the causes and policy implications of price volatility in food and agricultural markets and similarly got mixed results, not least because of the political nature of such studies (Interagency Report to the G-20 2011; Fonseca Blanco et al. 2010).

Price Dependent Regressions

Another genre in the literature is the reduced-form econometric analysis using quarterly or annual data (unlike the high-frequency time series data used in the econometrics literature surveyed in Chapters 7 and 8) that falls into the trap identified by Taleb (2012, 15):

> Experience is devoid of the cherry-picking that we find in studies, particularly those called "observational," ones in which the researcher finds past patterns, and thanks to the sheer amount of data, can therefore fall into the trap of an invented narrative.

These reduced-form time series econometric models typically have as regressors stocks to use ratio, crude oil prices, the exchange rate, interest rates, GDP growth, and inflation.

Baffes (2013) is associated with a number of papers, touting the "perfect storm" and showing that the storm is still alive and well (Baffes and Haniotis 2010). A sampling of these is described in Table 5.1, reflecting traditional economic analysis such as the effects of exchange rates, macroeconomic policies and shocks, speculation, commodity supply and demand trends and shocks, and the behavior of stockholders. Baffes (2013) argues that, because agriculture is so energy-intensive, crude oil prices have played a key role in commodity prices as input costs have risen. The transmission elasticity from energy to agricultural commodity prices is estimated to be 0.20 for the period 1960–2005 and considerably higher when the post-2005 boom years were included in the analysis. However, these increases in elasticities are present in all commodity sectors, thus arguing common factors are at play.

Nevertheless, the empirical results in Baffes and Dennis (2013) show just three variables are significant in explaining food commodity prices: stocks to use ratio, the crude oil price, and the exchange rate. These can as easily be interpreted to be the impact of biofuel policies, not through input cost effects but through price linkage effects, with the stocks to use ratio reflecting the low stocks due to the biofuel demand shock in the first place. More recently, Baffes and Etienne (2014) conclude in their abstract "other key drivers of food prices include (in order of importance) energy costs, physical stocks, and monetary conditions."

Gilbert (2010, 388) is also reflective of this literature:

> Agricultural price booms are better explained by common factors than by market-specific factors…A capital asset pricing model-type model shows why one should expect this and Granger causality analysis establishes the role of demand growth, monetary expansion and exchange rate movements in explaining price movements

He goes on:

> The demand for grains and oilseeds as biofuel feed stocks has been cited as the main cause of the price rise, but there is little direct evidence for this contention. Instead, index based investment in agricultural futures markets is seen as the major channel through which macroeconomic and monetary factors generated the 2007–2008 food price rises.

There are a number of empirical studies focusing on macroeconomic factors. A representative paper of this vast literature is that of Frankel and Rose (2009). Overshooting,[12] excess liquidity, GDP growth, fiscal expansion, portfolio diversification into commodities, and an emphasis on the theoretical link between real interest rates and real commodity prices are considered to be overriding factors affecting commodity prices. Frankel and Rose (2009) note the sub-prime mortgage crisis occurred in August 2007 and global growth forecasts fell after that but commodity prices not only did not decline but accelerated after that. Hence, the importance of interest rates, volatility in the spot-futures spread and "normal backwardation," inventories, bandwagon effects, and political, financial and economic risks (e.g., of a supply disruption).

How Each "Perfect Storm" Factor Relates to Our Theory for the 2008 Price Boom

Selective empirical evidence in support of the "perfect storm" explanation will be balanced with questions on the disputed role of several factors, including cases of observational equivalence, indicators rather than causal effects, different counterfactuals, and different approaches. We highlight the following "perfect storm" factors in relation to our theory: speculation, developing country policy response, exchange rates, supply and demand shifts, declining public R&D, and macroeconomic factors. However, before we do so, it is instructive to first remind ourselves of the literature we place ourselves in.

Our Theory of Biofuel Policy

The literature we are trying to establish argues that biofuels policies play a much bigger role, using specialized models that take a close look at which biofuel policy is binding and the specific relations between the gasoline (diesel), biofuel, and feedstock or crop prices domestically and internationally.[13] This literature focuses on the new and unique role of energy and environmental policies that created a direct link between biofuel and crop prices. Biofuels policies are varied: biofuel consumption mandates; biofuel consumption subsidies (e.g., tax exemptions); production subsidies for both biofuels and feedstocks; environmental regulations; import tariffs and tariff-rate quotas; and binary sustainability standards requiring biofuels to reduce greenhouse gases relative to fossil fuel. Each of these biofuel policy categories has its unique impact on grain prices—some long run, some fleeting, and some large effects, especially in combination with other biofuel or agricultural policies within country or across countries.[14]

We call our theory "the theory of biofuel policies" for want of a better term. This theory argues that most of the price increase was due to biofuel policies. We issued four widely distributed working papers in 2007 near the very beginning of the crisis (subsequently published in 2008 and 2009) that were widely ignored at the time (save Mitchell and Collins). Table 2 in de Gorter and Just (2007d; 2008b), for example, shows that the ethanol price premium due to policies was 65 cents a gallon in 2007–2008. The corn–ethanol price multiplier was estimated to be about four. This means corn prices increased $4 per bushel for every $1 increase in ethanol prices due to ethanol policies for a total change of 120 percent. In de Gorter and Just (2007d; 2008b), we argued that this high level of change appeared impossible because it ignored the important concept of "water" in the ethanol price premium due to policy. These concepts were used in a string of papers evaluating different issue regarding biofuels, including welfare effects, interaction effects between mandates and subsidies, carbon leakage and indirect land use change.

The approach departs from the standard literature of commodity price determination by modeling the explicit relationships between corn and ethanol prices, and between ethanol and gasoline prices, and the economics of blend mandates. This literature was fortified with papers by Drabik (2011), Cui et al. (2011), and Lapan and Moschini (2012) yet scant attention was given to them outside of the few who recognized the importance of the new price linkages. This ignorance continues to this day. Wright (2014) and Bobenrieth et al. (2014) do not mention this literature when expounding the importance of storage for the price run up. Nor does the series of papers by Abbott et al., written for the Farm Foundation, use our modeling framework. Likewise, the econometric literature on price links ignores our models—yet, the implications of our model are critically central to the understanding of price links and how to determine if biofuels have an influence or not. The less of a link between corn and crude oil prices (even negative), the higher the impact of biofuels on crop prices, because it indicates the mandate is binding with ethanol price premiums above the tax credit! The literature on explaining crop prices reflects a phenomenon economists suffer from, as so eloquently written by Taleb (2012, 9):

> That category of person who...defaults to thinking that what he doesn't see is not there, or what he does not understand does not exist. At the core, he tends to mistake the unknown for the nonexistent

We now briefly comment on how each of these "perfect storm" factors identified in the literature relate to our framework of analysis. We take 2008 as the example, but it extends to other time periods. However, later time periods require some greater nuance in analysis because the mandate premiums over the tax credit in these time periods are sometimes substantial, especially after the US blender's tax credit expired.

Speculation and Bubbles

Several economic studies cited earlier and international organizations (see Robles et al. 2009; De Schutter 2010) have declared speculation by new financial index investors either created or were significant contributors to the price booms. However, we showed coarse grain and oilseed prices were locked onto crude oil prices all of the way up and all of the way down in 2008. How could speculation have had any impact? Many economists have studied speculation in great detail to show otherwise. The most prominent is Scott Irwin at the University of Illinois who has published study after study with the biggest econometric models and data sets with the state-of-the-art econometric and statistical modeling techniques to show that it was not speculation (e.g., Irwin and Sanders 2011).[15] They have corn soybeans, soybean oil, feeder cattle, live cattle, and lean hogs in their data set. We can prove the first three commodities were not in a speculative mode just from our analysis of price links, using our theory. On the other hand, our theory cannot prove there was no speculation in the lean hog market; therefore, Irwin's countless studies are clearly worthwhile.

Exchange Rates

There are two basic points that we need to keep in mind. First, exchange rate movements since September 2006 have not been that big. Second, the incidence of exchange rates, like developing country policy responses for 2008 discussed in the next chapter, is zero when corn prices are locked onto crude oil prices. The only way exchange rates can impact food commodity prices in 2008 is if it affects crude oil prices. However, crude oil prices are often the cause of exchange rate movements, and if one applies Hotelling's theory of exhaustible resources, devaluation is a tax on crude oil prices and, therefore, oil producers respond by producing more oil, thereby offsetting the devaluation (Sinn 2008). We find it remarkable that the profession locks itself onto Marshallian supply and demand curves and stockholding behavior and are reluctant to let go and embrace other frameworks of analysis that are so obviously central to food commodity price determination in this new biofuel era.

Supply Shifts

The effects of bad weather have pretty well been discredited as an explanation as the variance in production of all crops has declined (see Figure 8 in Wright 2014) and, given the law of one relative international price for grains and oilseeds, it cannot be an explanatory factor for the commodity price boom that first became visual in September 2006. Furthermore, crop yields in Africa have increased.[16]

Nevertheless, many economists argue the wheat production shortfall in 2007–2008 was a primary contributor to the wheat price increase. We plot the US Department of Agriculture monthly forecasts of total world wheat production, spaced by crop years, in Figure 5.5 and find it incredible that one can detect a significant change in wheat production through the years that would justify the wheat price boom in 2008. Of course, the arguments hinge around inadequate stocks or increasing stocks, a topic we will discuss in more detail in Chapter 8.

Asian Demand Growth

Many economists have agreed that India and China were net grain exporters throughout the time period and, thus, cannot be a significant contributor to the price boom. Baffes and Etienne (2014) provide an excellent discussion of the literature and provide evidence themselves that Asian demand growth was not only inconsequential, but that one would expect a negative relationship between demand growth and grain prices.

Declining Public R&D

Many commentators remark that the price boom was comeuppance for declining public R&D in agriculture. Ball et al. (2013, 436) state "Alston, Beddow, and Pardey (2009), for instance, suggest that declining crop yields are the proximate cause of recent food price increases." However, we do not think the lack of R&D caused the price boom as the market was well aware of developments (e.g., because of the excellent works of Alston et al. 2010) and, therefore, it was not a surprise, unlike biofuels and the crop–biofuel price link that was suddenly created.

The price boom was not only a surprise, but it sobered up domestic governments and international organizations to reverse the long downward trend in public R&D for agriculture. Guariso et al. (2014) show how high prices heightened public awareness of global hunger and poverty and so galvanized domestic and especially international agencies to vastly increase funding for R&D in agricultural production, and for agricultural development and food security.

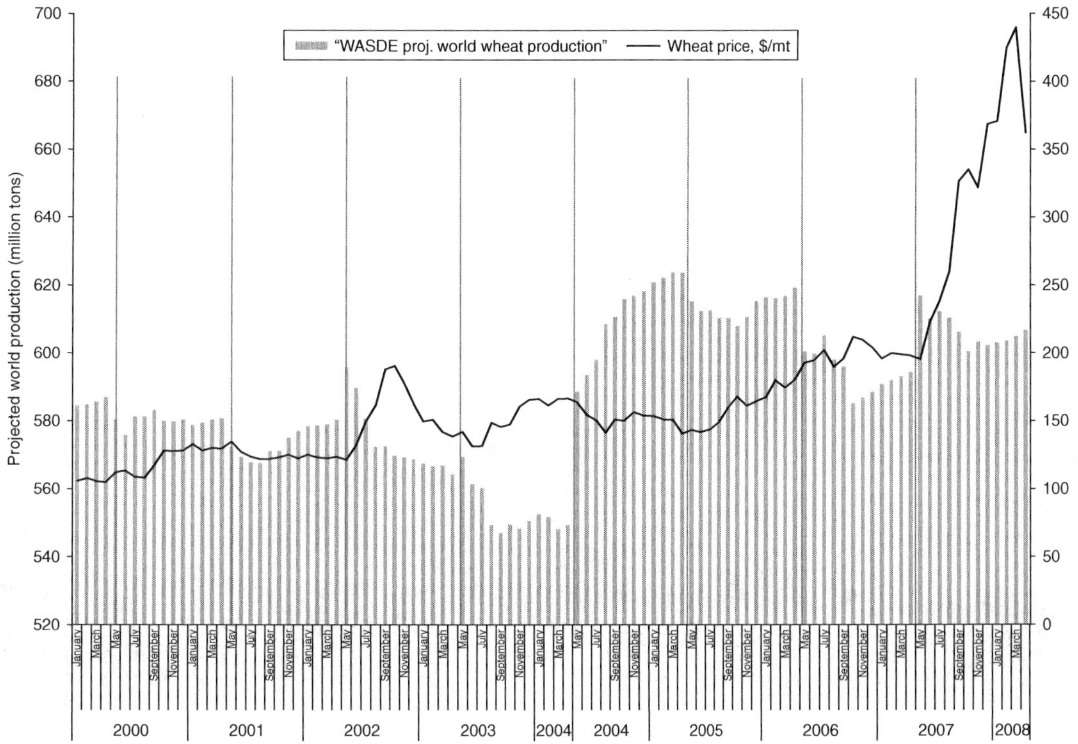

Figure 5.5 Wheat price boom due to bad weather in 2007–2008?

Source: WASDE (various issues) Projected total world wheat production, USDA Wheat Yearbook Table 20.

Macroeconomics

The problem with all of these studies putting top priority on macroeconomic influences (described earlier in this chapter) is that they do not recognize that this time is different. Shiller (2008) emphasizes that one of the primary drivers of asset bubbles is the notion that, somehow, this price spike is different from all others. This may be true in the boom equity market of 2001, or the boom real estate market of 2008. However, agricultural commodities markets actually transformed in 2006. Failing to recognize a fundamental change in markets can be just as damaging as believing in phantom changes. Supposed "asset bubbles" recently have only occurred in farm land prices; not housing or other asset classes. But it is not a bubble: it is grounded in the fundamentals—biofuels. There has been no general inflation. Interest rates have been close to zero. The discussion of Figure 1.3 emphasized how crude oil prices led cereals this time (crude oil followed cereals between the years 2002 and 2004). This is a big difference as the crop–biofuel price link was established this time (because of biofuel policies). In fact, there are very few characteristics of the 2006–present boom and bust (or rather boom–bust–boom again) in explaining food commodity prices that are similar compared to the period 1972–1974 (and those similarities are mostly irrelevant).

Concluding Remarks

Our approach is to determine how much of the price increase was due to the corn (soybean) price link to crude oil prices (which were lowest when locked). If prices were at this level, as it was all of the way up and all of the way down, then we can back out the effects of quantity changes. In this case, the effects of the other "perfect storm' factors were non-existent in 2008 and, on average, constituted a maximum of 20 percent of the action over a four-year time period.

Headey and Fan (2010) emphasize their "balanced" and "unusually comprehensive assessment of the crisis." What is of great concern to us is the overuse of "complexity" and "perfect storm" factors in all of these papers. Further, they suggest that if we conclude that biofuels is the most important factor by far in explaining the 2008 price spike and those thereafter, is our analysis then "unbalanced" and not "comprehensive"?

Another perplexing omission or remarkable glaring hole in the literature by economists studying the impact of biofuels is that feed grains and oilseeds were locked onto crude oil prices all of the way up and all the way down in 2008. It had no choice and did not care what happened in wheat and rice markets. But, the reverse is not true; what happens to 62 percent of the world's crop land has a big impact on wheat and rice markets: competition for land and substitution in demand. This asymmetry, once realized, casts a different shadow on what happened in 2007–2008.

Are we experiencing the emergence of a structural transformation due to biofuels or just deviations from longer term trends? When is a "perfect storm," which goes on for eight years, no longer considered a storm? Or is it so perfect that it goes on forever? When are a boom and a bust over? Or has it been a boom–bust–boom? Or just boom? Or a never ending boom and bust?

CHAPTER 6

THE ECONOMICS OF DEVELOPING COUNTRY POLICY RESPONSES AND BIOFUEL POLICIES

Recall in Chapter 1 that corn prices almost doubled in the six months beginning in September 2006, precipitating the Mexican tortilla crisis in January 2007 and India's ban of wheat exports the following month. This was the beginning of a long list of countries restricting exports (export bans, export taxes, value-added tax rebates, and actions by state trading enterprises and government to government sales) and promoting imports (lowering import barriers and manipulating domestic prices to be below world prices).[1] Even peasants hoarded rice (Timmer 2008), which can have the same effects as a developing country policy response. Developing countries are presumed to be responding to a world market price shock, which we argue was due to biofuel policies whereas most others argue it was a "perfect storm" of factors. The source of the shock does not concern us here; it is the economics of the policy responses.

The overwhelming consensus is that these policy responses by developing countries had a significant impact in not only increasing world prices in 2007–2008 and thereafter but also in increasing world price volatility. Furthermore, the literature emphasizes the self-defeating effects of developing country policy responses: as more exporters restrict exports and importers encourage more imports, world prices increase even further, thereby raising domestic prices in the same countries imposing the policy responses to protect domestic consumers. Of course, these policy responses hurt poor consumers in countries that do not have policies to insulate their domestic prices from world price increases.

All studies on the causes of the 2007–2008 food commodity price spikes are in agreement that developing country policy responses were a significant contributor to higher prices and price volatility (even in our own past writings, and as we will show in this chapter, like all others, are probably in error). Keep in mind that although most studies cite a long list of factors as causing the 2007–2008 price spike (mostly understating biofuel policy), there is a significant amount of disagreement across studies on the contribution of particular singular factors. For example, several studies conclude speculation had little impact (e.g., Abbott et al. 2008). Others, for example, conclude Asian income growth had little to do with it (e.g., Baffes and Etienne 2014). However, no study has ever disputed the important role of developing country policy response.

Therefore, the purpose of this chapter is to take a close look at the implications of our "theory of biofuel policy" in regard to the effects of developing country policy responses. All empirical studies that analyze the impact of developing country policy responses are for the year 2008. The most cited evidence is given by Anderson et al. (2014a): the price of rice, soybeans, wheat, and corn rose 52, 31, 18, and 18 percent, respectively, and due solely to developing countries restricting exports or promoting imports.[2] It should also be pointed out that most studies analyzing developing country policy response are analyzing the impacts on poverty in developing countries. We will show that developing country policy responses had little impact on world prices and maximum impact in reducing domestic price in developing countries. Thus, this chapter is an outlier in that we digress and analyze the implications of our own framework of analysis for policy analysis. We vouched earlier that we do not do prescriptive analysis in this book. We digress here but just for a moment. We do not analyze the implications of our framework for all other explanations of the "perfect storm" identified in Table 5.1—that would be a book onto itself. However, this chapter does highlight the importance of our framework of analysis and the need for economists to adopt it in undertaking policy analysis in all of its dimensions, given this new biofuel era.

The Economics of the "Standing Up in the Stadium" Effect

Let us first present a simplistic model to explain the logic of the self-defeating effects of developing country policy responses that are essentially measures to insulate domestic market prices from world prices while at the same time exacerbating world price increases and volatility. Using traditional supply/demand analysis, Anderson et al. (2014a) argue that if many countries insulate by imposing export restrictions or adopting policies that encourage imports, the net effect is to increase world prices and the net effect may be zero insulation from the initial exogenous shock. Their argument is best summarized by Anderson et al. (2014b):

> Indeed, we have shown that if the same proportion of the world's exporting and importing country groups happened to insulate to the same extent, domestic prices in both country groups would rise just as much as if no country had insulated. This is akin to a crowd of people of equal height in a stadium trying to see better by standing up.

Therefore, both the world price increasing effects of developing country policy responses to an exogenous world price shock and the self-defeating aspects of developing countries trying to insulate their domestic prices from higher world prices is summarized in the term "standing up in the stadium effect."[3] We now explain this phenomenon using traditional supply and demand analysis, ignoring for the moment the theory of biofuel policy as developed so far in this book.

Consider an exogenous shift left in the world's excess supply of grains and oilseeds due to biofuels policies in rich countries diverting grains and oilseeds to biofuels. This is depicted in Figure 6.1 as a shift in the excess supply curve ES to ES′, causing world prices to increase from P_{W0} to P_{W1}. Because of import tariffs, the domestic price in developing country importers increases from P_{M0} to P_{M1} due to the shock in excess supply (from point *a* to point *b* in Figure 6.1).

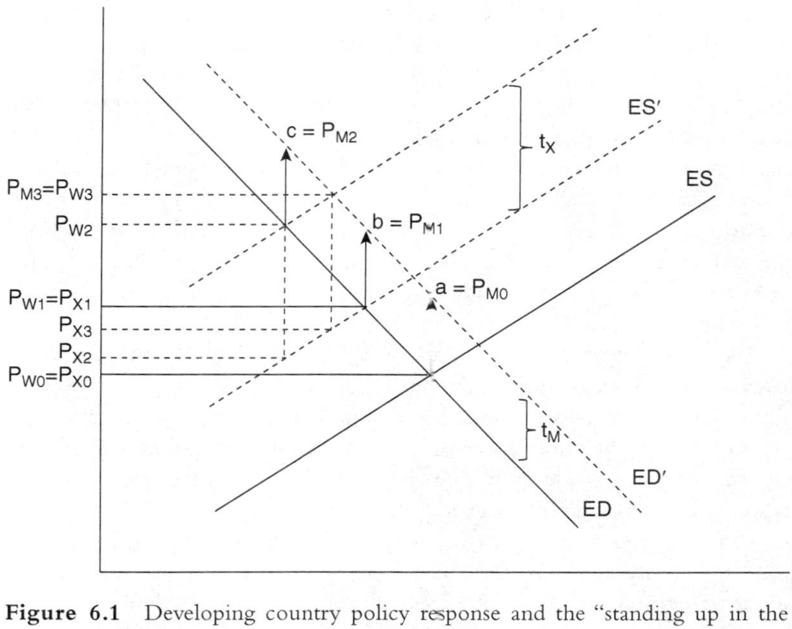

Figure 6.1 Developing country policy response and the "standing up in the stadium" effect.

Now consider an export tax of t_X by developing countries. The world price increases to P_{W2}, domestic prices in developing countries that impose the export tax declines to P_{X2} and domestic prices in developing country importers with a tariff t_M rises to P_{M2} (point *c*). Finally, importing countries eliminate their tariff and the final equilibrium world price is P_{W3} (equal to domestic prices in importing countries) and the domestic price in exporting countries is P_{X3}. In this example, domestic prices in exporting countries that try to insulate from world price shocks are lower than the post-shock price of P_{X1}, but this need not be; it depends on the relative shifts in excess supply and demand due to policy responses and the relative elasticities of excess supply and demand. However, the weighted average domestic prices are likely to be higher than P_{W1}; therefore, it has been self-defeating. Furthermore, world prices are higher due to developing country policy response, moving from P_{W1} to P_{W3}.

The Economics of Biofuel Policies and Developing Country Policy Responses in 2008

But, how could such policy responses to higher prices have any impact on the prices of soybeans and corn when they were locked onto the crude oil prices all the way up and all the way down in the 2008 price spike? Coarse grains and oilseeds represent 62 percent of all cropland. And if wheat is easily substituted for corn as feed, why would we not expect wheat price to keep pace as corn prices rise? Recall that white corn (used to produce tortillas consumed in Mexico) rose 107 percent in the six months after September 2006—US farmers began to use white corn for feed as it is not used in ethanol production. And if wheat prices triple as they did in 2007–2008, why would we not expect rice prices to go up due to similar substitutions? More wheat is consumed in Indonesia than rice and Asians are quite capable of switching. This is especially the case when a couple billion people spend close to 50 percent or more of their income for basic foodstuffs.

Therefore, let us characterize a world grain and oilseed market in 2008 as one that is locked onto crude oil prices (through the appropriate price links discussed for corn—applicable to wheat as well—and soybeans). What would one expect the impact of developing country policy response to be? Consider the world price of a crop to be locked on the crude oil price through the tax credit. The impact of an export restriction in response to an exogenous increase in the world price is given in panel (a) of Figure 6.2. The excess supply curve for the exporting country is given by ES. The exporting country faces a flat excess demand curve ED (which represents the world price). The exporter could be a "large country" exporter in the

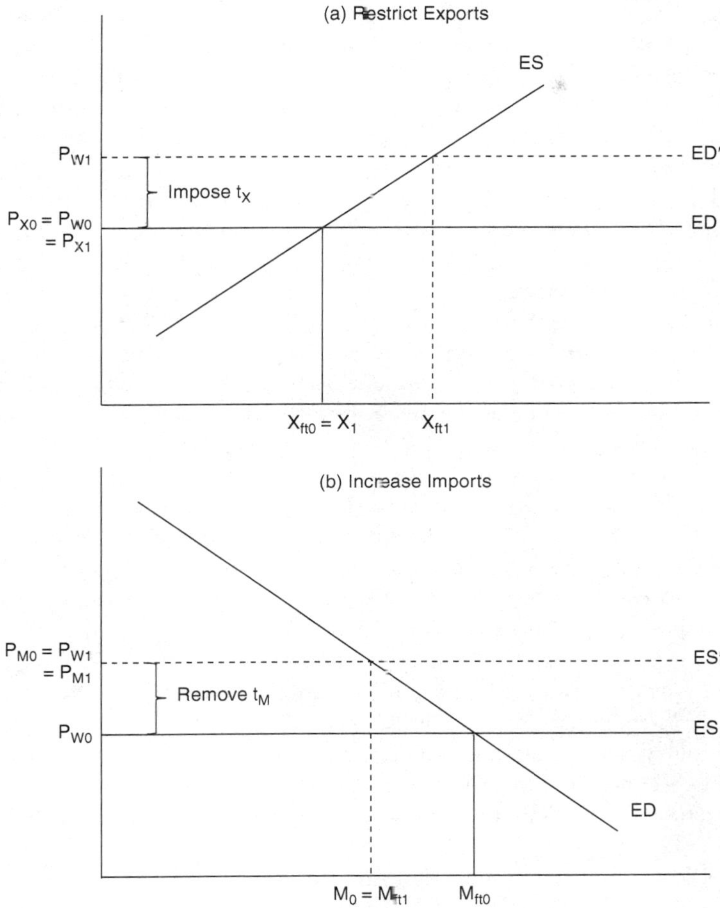

Figure 6.2 Effects of developing country policy response in 2008 assuming biofuel price links.

traditional sense but here faces a perfectly elastic excess demand curve (because we assume the *change* in biofuel production due to developing country policy response does not impact crude oil prices, an assumption when relaxed will make little empirical difference in our results).

We depict the export tax equivalent of the export restriction by t_X which is shown to exactly offset the exogenous increase in price (this is not necessary; obviously, some countries partially offset the world price increases; other countries may have overcompensated). Because we have selected a restriction that exactly offsets the exogenous price increase in panel (a) of Figure 6.2, exports and, therefore, domestic producer and consumer prices remain unchanged from the levels observed before the world price spike. The incidence of the policy is a 100-percent reduction in the domestic market price spike compared to what it would have been without the policy.

The exact same story can be given for an importer that reduces import barriers. The excess demand curve ED is given in panel (b) of Figure 6.2, where the (perhaps "large country") importer faces a perfectly elastic excess supply curve that represents the fixed price of the grain or oilseed (because it is locked onto the world crude oil price). The effect of an importer reducing import barriers is given in panel (b) of Figure 6.2. Again, we arbitrarily show a full offset of the exogenous increase in the crop price by promoting imports by the import tariff equivalent t_M and the incidence is once again 100 percent in reducing the domestic price spike compared to if no actions were taken.

The results of our characterization of the market is in sharp contrast to that of the literature (e.g., Anderson et al. 2014a) who argue government policy responses not only exacerbate the international price spike but also are self-defeating in terms of insulating themselves from the original world price shock (as per our discussion of Figure 6.1).

Nevertheless, our (albeit simple) characterization of the market and impacts of policy response in Figure 6.2 gives a very different picture. There is little "standing up in the stadium" effect. Figure 6.2 by construction has developing country responses of the eliminated import tariff (by importers) and export tax (by exporters) maintain domestic prices at pre-shock levels in each case. There is no effect on world prices because corn (and, therefore, all grain/oilseed) prices are locked onto crude oil prices (the impact of the change in biofuel production on crude oil prices is ignored to keep the analysis simple). Therefore, consumers are fully protected from price increases. Producers are unable to benefit from price increases.

Taking the case of corn, the economic impact of developing country policy response is to shift out the non-ethanol demand curve for corn (D_{NE} in Figure 3.1 of Chapter 3) and, therefore, to reduce the amount of corn going into ethanol. The lower domestic supply and more domestic demand due to developing country policy responses means less grains and oilseeds can be diverted to biofuels in the rest of the world. The reduced domestic supply and increased domestic consumption of grains and oilseeds in developing countries imposing export restrictions and import enhancements reduce the supply to the rest of the world but crop prices are locked onto crude oil prices and, thus, are unaffected. Biofuels consumption declines. In theory, all of the increased consumption and reduced production in developing countries due to policy responses is crop that otherwise would have gone to biofuel consumption but now is eaten by consumers, including poor consumers. This is what would happen, unless the mandate is binding. If the mandate is binding, the reduction in grain and oilseed available for biofuels drives the world price up even further as blenders scramble to meet a less and less attainable mandate. However, that is not the avenue by which Anderson et al. (2014a) argue developing country policy responses operate. And the mandate was not binding in 2008, the year all studies on the economic impact of developing country policy responses analyze.

The consumption-weighted average price of cereals in developing countries for 2008 is given in Figure 6.3. The weighted average domestic price in developing countries did go up in 2008, but that can be explained by various factors consistent with our hypothesis. It seems like the policy reactions were effective in containing the increase in the domestic price. Our model can explain this lack of price response. Some countries may have overreacted (in other words, t_X or t_M were greater than that depicted in Figure 6.2) or underreacted. Overall, according to our theory, given the empirical evidence for 2008 in Figure 6.3, countries slightly underreacted. Baltzer (2013) shows India overreacted for wheat (domestic wheat prices

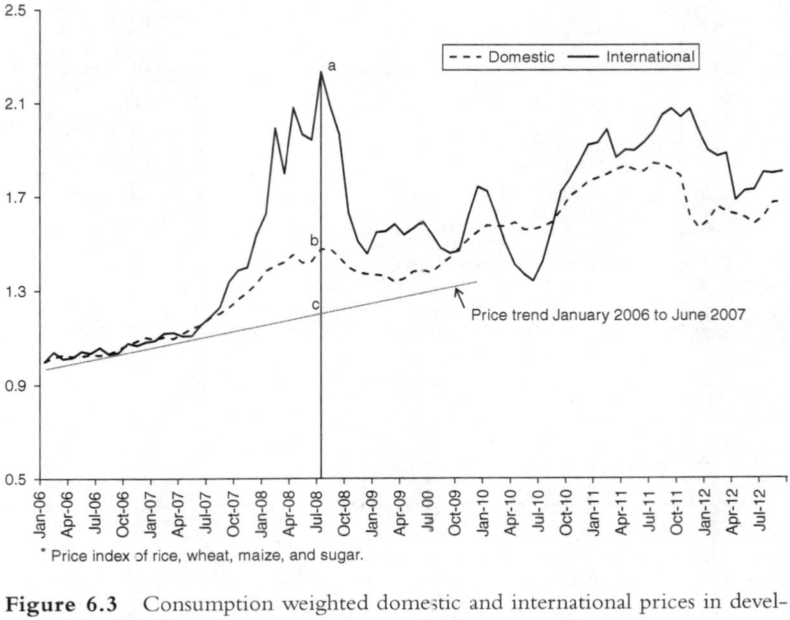

Figure 6.3 Consumption weighted domestic and international prices in developing countries.*

Source: Ivanic and Martin (2014b).

declined in 2008) while in China, there was a slight increase. China left soybean prices to follow world markets. Therefore, an outcome in Figure 6.3 is entirely plausible with our thesis. For rice, Bangladesh, China, and Indonesia, among others, kept prices low through their policy measures. For wheat, China, India, Turkey, Pakistan, and Japan were among the countries that were able to keep domestic price below the no country-intervention level with their policy reactions.

Table 6.1 uses the average monthly data for 2008 from Figure 6.3, with the baseline world price taken from the trend before the price boom. Therefore, the first four rows of Table 6.1 summarize the baseline data that is the same for both models. Weighting the Anderson et al. (2014a) estimates of the price changes in each crop by consumption of calories, we estimate the price increase due to developing country policy response implied by Anderson et al. (2014a) to be 28.75 percent. From this, we calculate the implied world price after the shock but, with no developing country policy response (in our analysis, we assume the world price does not change with the policy response). We can then calculate the alleged "standing up effect" of 0.178 and compare this to the actual outcome which is −0.259. In fact, the final row gives the share of the price increase due to the initial

Table 6.1 Standing up in the stadium effect of developing country policy responses (average 2008)

	Anderson et al. (2014a, b)	*Model depicted in Figure 6.2*
Final world price (pt *a* in Figure 6.3)	1.859	1.859
Initial world price (pt *c* in Figure 6.3)	1.24	1.24
Change in world price (pt *a* minus pt *c* in Figure 6.3)	0.619	0.619
Domestic price developing countries (pt *b* in Figure 6.3)	1.422	1.422
World price after shock	1.681[a]	1.859[b]
Standing up effect		
Alleged	0.178	0
Actual	−0.259	0
Percentage of price increase due to shock closed because of policy responses	59	71

Notes:

[a] Excluding price increase due to developing country policy response of 28.75 percent using consumption shares for wheat, corn, rice, and soybeans with price increasing effects of 51, 18, 18, and 31 percent, respectively (Anderson et al. 2014a).

[b] Assumes no impact on world prices of developing country policy response as per Figure 6.2.

Source: Calculated.

shock that was closed by the policy responses. The data do not support the "standing up effect" when we combine the information in Anderson et al. (2014a) with that in Figure 6.3 derived by Ivanic and Martin (2014b). The difference between our simple models in generating 71-percent effectiveness with that of theirs (59 percent) is not that large and is acknowledged in Ivanic and Martin (2014b), who recognize the price shock was significantly dampened by developing country policy response.

The evidence appears to be in favor of our hypothesis that the incidence was closer to 100 percent in the domestic market than a standing up in the stadium effect. We maintain that for 2007–2008, the world prices for 62 percent of the cropland could not have been affected at all by these developing country policies. The only remaining question is how strong the law of one relative price of wheat vis-à-vis coarse grains and oilseeds or rice must have been?

Jones and Kwiecinski (2010) provide evidence that China, India, and Ukraine reduced wheat exports significantly as did China and Ukraine for corn, and China and India for rice. Dawe and Timmer (2012, 129) underline how:

> During the world rice crisis of 2008, three countries (China, India and Indonesia, the three most populous developing countries in the world) successfully insulated their domestic rice economies from the turmoil on world markets....this was one reason why the crisis pushed fewer people into poverty and undernourishment than was initially feared.

This is a conclusion entirely in line with our theory of biofuel policy. Anderson et al. (2014a) ignored the impact biofuel policies may have had. In fact, the word "biofuels" appears in another earlier paper (if only briefly) by Ivanic and Martin (2014a, 273):

> Other sources of volatility that have been emphasized include government programs such as biofuel mandates, which link the price of grains to those of volatile fuel prices; however the evidence regarding the impact of such mandates on global price volatility remain inconclusive. (Zhang et al. 2010)

In other words, Anderson et al. (2014a) absolve themselves of any responsibility for recognizing the role of biofuel policies and, thereby, require no adjustment to their traditional form of analysis by relying on the oft-cited paper by Zhang et al. (2010) that, like many in this genre (to be reviewed in the next chapter), do not find a link between crop and biofuel prices and, therefore, (erroneously) conclude biofuel policy is not of relevance.[4] Thus, they do not consider the possibility that crop prices are related to energy prices when there are no mandate premiums, as was the case in 2008. Anderson et al. (2014a) were influenced by a literature that is completely misplaced, and by one particularly overhyped paper that also reads:

> The underlying hypothesis states rising biofuel prices would cause a shifting of agricultural commodities from food toward fuel production, which should then drive up agricultural prices. However...time-series results do not tend to support such a hypothesis...Rising sugar prices appear to be the leading cause of price inflation of agricultural commodities. Zhang et al. (2010, 449)

Notice some key phrases: "Biofuel prices...do not drive up agricultural prices"!? And that "rising sugar prices appear to be the leading cause of price inflation of agricultural commodities"! Figure 6.4 shows otherwise. Figure 6.4 kills three birds with one stone for us.[5] First, Zhang et al.'s (2010) thesis that sugar prices drove commodity prices in 2008 has little basis in fact. Second, several economists have remarked over the years that all commodities followed crude oil prices and our theory has little meaning. The data in Figure 6.4 indicates otherwise. Finally, for those who think speculation was so important in causing the price boom, why was there speculation in rice and not in cotton or sugar markets?

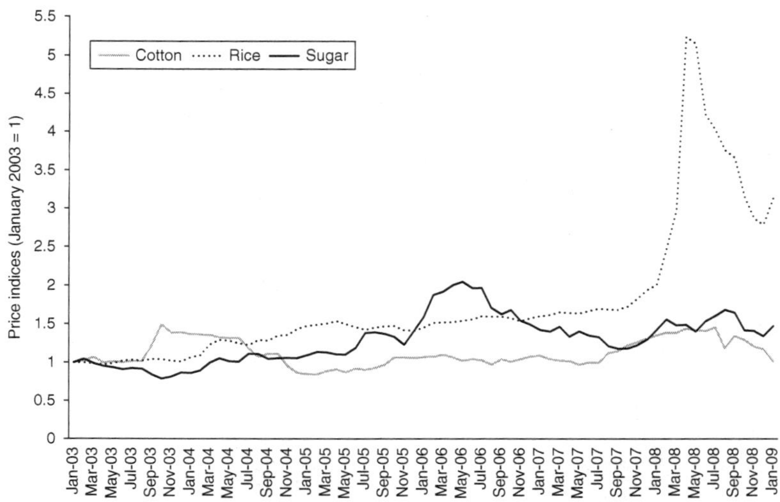

Figure 6.4 White commodity prices: Greedy speculators causing "euphoric bubble"? Or all commodities followed oil prices? Or sugar prices led the way?

Source: Calculated from monthly data in World Bank (2014).

Concluding Remarks

As an example of our many writings on biofuel policies and prices, de Gorter et al. (2013e) state:

> Many developing country policy responses which may also have contributed to higher prices…The policy responses by developing countries to higher commodity prices in 2008 in the form of export taxes or reduced import barriers exacerbated the international grain/oilseed price volatility even further.

For the 2007–2008 price boom, we are probably in error. Perhaps we should get a pass because we have not formally studied the impact of developing country policy responses. Yet, we study the impact of biofuel policies closely so we should have understood the implications of our own theories and price linkages such that developing country policy responses are expected to have had a small impact in 2008. Therefore, maybe we should simply declare that the profession may have made an error in regards to the issue of the effects of developing country policy responses. This is significant because many studies on developing country policy response analyze the implications on poverty. The implications maybe be altered significantly in that respect; the quote by Dawe and Timmer (2012) above or the analysis by Galtier (2013) may be diagnosing the issue more accurately than most of the literature that concludes otherwise.

Using our theory of biofuel policies in analyzing the economics of developing country policy responses exemplifies the potential erroneous conclusions one reaches if traditional supply/demand analysis is pursued and our theory is ignored. The policy implications of the theory we develop in this book may have diametrically opposite conclusions. Therefore, all topics and factors identified in the "perfect storm" of factors explaining the price boom (summarized in Table 5.1) should be analyzed accordingly, but we leave that task for the future.[6] Nevertheless, this example highlights the potential importance of the framework of analysis presented in this book.

CHAPTER 7

AN ENHANCED EXPOSITION OF THE CORN–ETHANOL–ENERGY PRICE LINKAGES AND IMPLICATIONS FOR TIME SERIES ANALYSIS OF BIOFUEL POLICY IMPACTS

The purpose of this chapter is to critically assess a burgeoning literature that uses high-frequency time series (HFT) econometric analysis (using daily, weekly, or monthly data) that frantically seeks a link between feedstock and biofuel prices (say corn and ethanol), biofuel and gasoline/diesel prices, or directly between feedstock (e.g., corn) and crude oil prices. All of these papers explicitly or implicitly state that their objective is to determine the impact of biofuel policies on food commodity prices, and that their results have policy implications. To them, the impact of biofuel policies simply depends on the size and significance of the *positive* relationship between these prices: feedstock, biofuel, energy, or crude oil prices. Granger causality tests are also undertaken (studies even find corn prices "cause" crude oil prices!) If no link is found or the evidence is mixed then biofuel policies' impact on crop prices is just that: none or unclear, respectively. All papers in this HFT econometric literature come to one of these three conclusions on biofuel policies' impact on commodity markets: yes, no, or maybe.

However, we have shown so far that if mandates are binding, we expect the ethanol (and, thus, corn prices) to be weakly *negatively* related to gasoline (and, thus, crude oil) prices, with supply and demand shocks having a role as long as the mandate premiums above the tax credit are maintained. On the other hand, if the tax credit is binding, then corn and crude oil prices are expected to be strongly positively related, and supply/demand shocks have no role, not even higher input costs (unless the change in ethanol production affects crude oil prices—a feedback effect that is bound to be small). The impact of crude oil prices on commodity prices is, therefore, regime-dependent. This distinction between the two policy regimes is very important because (1) the causality between corn and ethanol prices dependents on the policy regime, and (2) the impact of supply/demand shocks on ethanol and hence corn prices also depends on the regime. Therefore, we summarize our theory and results in previous chapters and relate them to this literature.

Firmly implanted in this HFT literature is a paper by Enders and Holt (2011, 5) who state that:

> Surprisingly, what is missing in much of the current debate on commodity food price inflation is legitimate empirical analysis.

The reader will find that our analytical and empirical analysis is legitimate and surprising (and structural, elegant, and relevant). The surprise is thus: this HFT econometric literature comes to the exactly opposite conclusion than they should: if one finds a strong positive relationship between crop and fuel prices, then the tax credit regime is binding (and in theory, corn prices are lowest when locked onto crude oil prices through the ethanol and gasoline prices). When this literature finds no relationship or a negative one, then mandate premiums exist and, therefore, biofuel policies have their maximal impact! Furthermore, when the HFT finds little relationship between crop and biofuel prices, as the discussion of our model's prediction errors in Figure 2.2 shows, inadequate attention may have been given to biofuel production, marketing, and blending constraints, each of which can break a price link between corn and ethanol, and between ethanol and gasoline. This does not mean our framework or biofuel policies are irrelevant; on the converse, one needs our framework to determine what is going on in the market.

We, therefore, take a critical look at the burgeoning HFT literature that analyzes the dynamic linkages among crop, biofuel, and energy prices using statistical time series models. The purpose of this literature is to determine the impact of biofuels on crop prices by measuring the influence of higher energy prices on crop prices. However, it is through the price linkages (or non-linkages in the case of mandate premiums where ethanol prices float up and away in the opposite direction of gasoline prices) that biofuel policies had its biggest effect. Ironically, we have, so far, emphasized the price-link approach. Quantity changes through crop supply/demand shocks only matter if there are biofuel price premiums due to mandates, and the extent to which such quantity shocks matter is up to the point such premiums still exist. Shocks in crude oil supply or demand (or crude oil prices) are another matter discussed in Chapter 8.

This HFT literature uses modeling techniques that are atheoretical time series models, woefully inadequate of being able to establish or quantify causal links in the data. Unlike the structural model we have developed so far, this HFT literature does not specify a structural econometric framework that acknowledges the underlying identification problems. We showed mandates and tax credit bounced back and forth over the historical time period. None of these papers even recognize that the statistical relationships they are seeking are regime-dependent; in fact, mandates versus tax credits are not even mentioned in this literature.[1]

We also showed there were periods with ethanol-blending constraints (ethanol was sold at a discount to gasoline) and ethanol-production constraints (ethanol and corn prices diverged from their equilibrium values). Any statistical analysis has to account for these constraints too, but it would require a rigorous structural model (that we have developed for corn–ethanol only so far) with the appropriate equations and predictions to determine when these disequilibria occur and what their magnitude is. We will also show that the source of shock(s) will matter in terms of the expected sign of price relationships. This HFT literature is fundamentally without a theoretical basis. Our general framework of analysis provides implications for the expected signs and magnitudes of the various price links (corn–ethanol; ethanol–gasoline; and gasoline–crude oil) given the two regimes, the type and source of shock, and the presence of blending or production constraints.

This chapter is outlined as follows. The next section summarizes the expected signs and magnitudes of the various price links given our corn–ethanol model. To put everything together, we show that any statistical analysis of any one price link is frivolous and certainly cannot ascertain the extent to which biofuel policies affected commodity prices. The section after that provides a brief review of the HFT econometric literature in terms of the statistical methods used, the very mixed messages it gives (the results are all over the place) and the resulting shortcomings. This HFT literature is shown to be built on false premises, a critique of which provides a nice summary of all that we have learned in this book so far. This highlights the importance of having a sound economic model before embarking on these types of HFT modeling exercises.[2] The final section concludes.

Crop–Biofuel–Energy–Crude Oil Price Relationships: A Review of Our Theory

We summarize all of the relevant price linkages and direct impacts of changes in input costs due to crude oil price changes and crop supply/demand shifts in Figure 7.1. Changes in the relationships can be affected

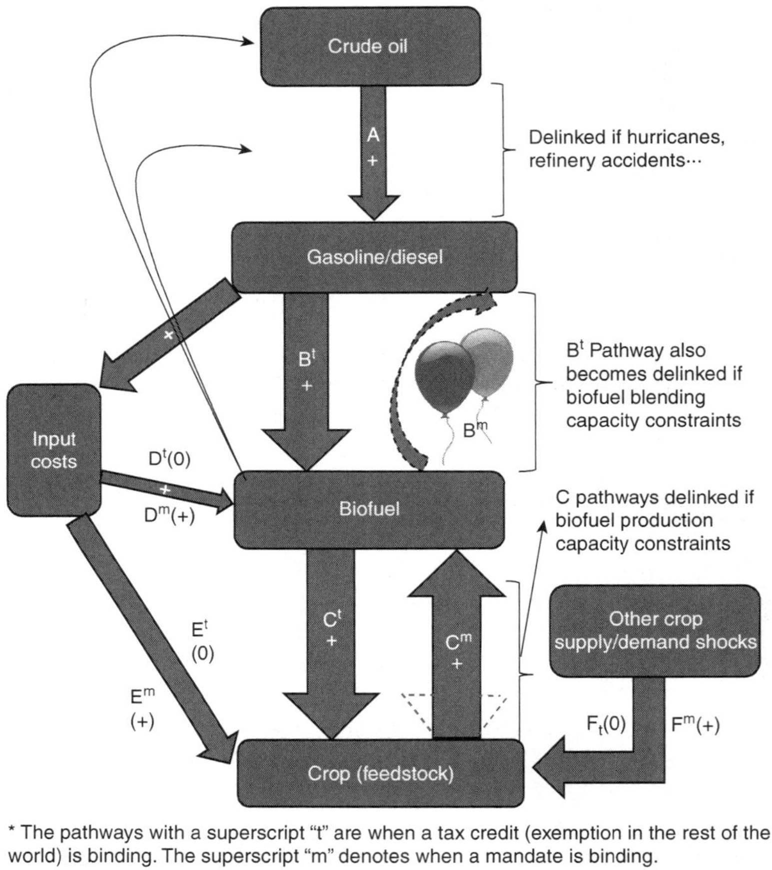

Figure 7.1 Energy–biofuel–crop price relationships.*

Source: de Gorter and Drabik (2014).

by changes in regimes and policies as well as prevailing market conditions. We use US corn–ethanol as an example in this chapter.

Crude Oil–Gasoline Price Link

Regardless of whether the mandate is binding or not, the relationship between the price of oil and the wholesale market price of gasoline remains the same: they are directly and positively related (Pathway A in Figure 7.1). A simple regression of the gasoline price on the crude oil price has a big error term, due to the omission of a number of technical parameters, including seasonality. However, the price link can also be broken due, for example, to hurricanes or accidents, thereby impacting gasoline and, thus, ethanol and corn price links and volatility. If one does not isolate these factors, and the literature does not, this is just the beginning of the errors that can affect empirical results in the HFT literature.

Does Ethanol Affect the Gasoline–Crude Oil Price Link?

There always is some impact of increased biofuel production on the level of crude oil prices (see arrow from biofuel prices back to crude oil prices in Figure 7.1). Using the empirical model of Drabik et al. (2010), gasoline prices decline by 5.1 cents per gallon under most plausible supply/demand estimates and 7.7 cents per gallon in a worst case scenario of vertical supply of oil[3] (5.7 and 8.7 cents per gallon with a mandate).[4] Cui et al. (2011) model the effects of ethanol production on lowering gasoline consumption and, thus, crude oil prices, but include the impact on petroleum by-products. When the gasoline price declines as a result of a biofuel policy, the value of a barrel of oil declines initially, other things equal, resulting in a lower level of crude oil and, therefore, petroleum by-product production. This reduction in production, in turn, raises the prices of petroleum by-products, placing a floor under the falling oil price. In reality, there is robust international trade in gasoline and by-products from oil processing such that biofuels probably have little effect on the oil–gasoline price relationship in the long run (Knittel and Smith 2015).

Du and Hayes (2012) argue the production of ethanol lowered the refiners processing margin between gasoline and crude oil, holding crude oil prices constant. This occurs because ethanol production causes excess capacity in crude oil refining, thereby reducing margins and market power. The estimated reductions in refiner margins by Du and Hayes (2012) were highly unrealistic, resulting in ethanol policy to reduce corn prices (see Box 3.1). Elam (2012) and Knittel and Smith (2014) find several other faults with this analysis.

Gasoline–Ethanol Price Link

If ethanol prices are locked onto gasoline and hence crude oil prices, an increase in oil price increases price of gasoline and, therefore, ethanol price increases as governed by equation (2.2) in Chapter 2. The Pathway B^t where the coefficient of price transmission depends on relative miles per gallon achieved of ethanol versus gasoline.[5] The gasoline price determines the ethanol price, with a significantly smaller feedback effect—shown in Figure 7.1 by the arrows back to crude oil prices—where increased ethanol production can lower gasoline and, therefore, oil prices. An increase in the oil price increases costs of processing ethanol (Pathway D^t where c_0 in equation (2.1) in Chapter 2 increases so corn prices decline through pathway C^t and C^m in Figure 7.1); however, the price of ethanol is locked onto the gasoline price and, therefore, ethanol prices do not change.

If the blend mandate is binding, an increase in the crude oil price increases prices of gasoline so the fuel (blend of gasoline and ethanol) price increases, causing both fuel consumption and, therefore, ethanol prices to decline (Pathway B^m).[6] This negative effect is not large when the blend ratio is approximately 10 percent. Ethanol-blending constraints cannot occur with the crude oil price declining in this situation. In this case, gasoline price has a negative effect on ethanol price. Equation (2.2) in Chapter 2 earlier does not hold: ethanol prices float up and away from gasoline prices (denoted by the balloons for Pathway B^m).

An increase in oil price increases costs of processing ethanol (Pathway D^m) as c_o in equation (2.1) of Chapter 2 increases and, therefore, the price of ethanol decreases relative to gasoline because a higher oil price means (1) a higher gasoline price; and (2) a higher gasoline price means a lower ethanol price under a binding mandate. This occurs independent of the corn–ethanol price link (unless one incorporates the feedback effect of ethanol production (prices) on gasoline/oil prices). But, one does have to analyze Pathways D^m and B^m simultaneously.

A shock in the fuel demand/gasoline supply will have a first-round effect in the gasoline market (the gasoline/diesel bubble in the diagram) and then gets transmitted as shown in Figure 7.1. Gasoline–ethanol price links in Pathway B^t can be broken by blending-capacity constraints, further affecting price links down the marketing chain depicted in Figure 7.1 (see discussion of Figure 3.2).

Ethanol–Corn Price Link

Regardless of what policy is binding, the corn–ethanol price link (Pathways C^t and C^m) is governed by equation (2.1). The causality is strictly positive from ethanol to corn prices for pathway C^t but mostly from corn to ethanol in Pathway C^m in Figure 7.1. The ethanol–corn price link can be broken by production capacity or ethanol marketing constraints (as described in Chapter 2, the latter occurring in 2013–2014). The impact on corn prices now depends on what policy is binding. If the tax credit is binding, then corn prices decline by the dollar per bushel equivalent of the excess profits in ethanol production. If the mandate is binding, then most of the incidence of removing excess profits is to reduce ethanol prices, rather than increasing corn prices. Nevertheless, capacity constraints affect the level and volatility of corn prices.

Crude Oil–Corn Price Link

If the corn price is locked onto crude oil prices (the tax credit binding as in 2008), an increase in the crude oil price will result in higher corn prices: Pathway A to B^t to C^t has corn prices increasing, but Pathway D^t has c_o increasing, causing corn prices to decline (the net effect on corn prices is most likely positive because higher oil prices affect ethanol production costs indirectly through natural gas and electricity prices). Of course, higher crude oil prices have no impact on corn prices through higher corn production costs if the tax credit is binding (Pathway E^t). Higher input costs for corn production with a binding tax credit shift the corn supply curve up and change the allocation of corn among uses, including a reduction in corn allocated to ethanol production as per Pathway E^t.[7] However, if the mandate is binding, then higher crude oil prices have a direct positive impact on corn prices (Pathway E^m), ceteris paribus, taking ethanol prices with them (Pathway C^m) (see explanation for Pathway B^m).[8]

Effects of Other Corn Supply/Demand Shocks on the Corn Price

Now let us take the case where the ethanol price premium over that under a tax credit is because the consumer has no choice between ethanol and gasoline; therefore, blenders do not have to price ethanol to gasoline strictly on a miles achieved basis but (perhaps partially) equal on a volume basis.[9] In this situation, the ethanol price is positively correlated with the gasoline price analogous to that described in equation (2.2) earlier. So, overall, the expected correlation between gasoline and ethanol prices is ambiguous and can, thus, in theory fluctuate between positive and negative values, depending on the situation. The expected negative relationship between ethanol and gasoline prices for a binding blend mandate holds when everything else is held constant. Given all the other factors shifting the various supply and demand curves over time, the positive correlation can exist although the mandate has mostly been binding, especially if some of the changes in gasoline price are due to demand shocks

Let us first take the case of a formal blend mandate binding (or equivalently, if environmental regulations require a particular percentage of ethanol in the fuel mix). If there is an exogenous increase in the gasoline price, then fuel demand declines along with ethanol prices. The ethanol and gasoline prices are expected to be negatively correlated by a coefficient greater or less than one. However, if there is an exogenous change in ethanol prices due to a change in the blend mandate requirement, then the correlation between ethanol and gasoline prices is ambiguous (see discussion in Chapter 2).

Table 7.1 Impacts of shocks under alternative biofuel policy regimes

	Tax credit *Effect on prices of:*		*Mandate* *Effect on prices of:*	
	Gasoline	*Ethanol/corn*	*Gasoline*	*Ethanol/corn*
Negative shock in ethanol (corn) supply	+	+	–	+
Negative shock in gasoline supply	+	+	+	–
Positive shock in fuel demand	+	+	+	+

Source: Drabik (2011).

If the tax credit is binding, any shift in corn supply or demand will only change the allocation of corn to different uses and corn prices are tied to gasoline (oil) prices: Pathway F^t (same reasoning as with Pathway E^t). If the mandate is binding, any shift in corn supply or demand will also affect corn prices, taking ethanol prices with them: Pathway F^m (same reasoning as with Pathway E^m).

The discussion so far highlights how the relationships among energy, biofuel, and feedstock prices are nuanced, but it is important to understand the impact of biofuels (and agricultural policies on grain and oilseed price levels and volatility). With a mandate binding, there are four negative effects and one positive effect of a higher oil price on corn prices.[10] Therefore, simply analyzing the crude oil–corn price link misses the complexities of the crop–biofuel–energy price linkages—many of the mechanisms or pathways have to be estimated directly, conditional on regimes (and if there are any capacity constraints, as emphasized earlier in analyzing the prediction error of our model in Chapter 2).

A summary of our results with different kinds of market shocks are given in Table 7.1. Under a tax credit, a negative shock in gasoline supply has the same effect on market prices as a positive shock of the same size in the fuel demand. Under a blend mandate, a negative shock in gasoline supply has always a greater effect on gasoline prices than a positive shock of the same size in the fuel demand. One implication of the results in Table 7.1 is that an oil price shock due to supply disruption and due to oil demand growth in Asia will reinforce each other's effect on corn prices when the tax credit is binding but weaken each other under the mandate.

Results from the HFT Literature on the Various Price Links

Although many studies espousing the perfect storm explanation of the recent price rise attribute only a small role to biofuels and biofuel policies, a growing body of literature has emerged that analyzes the dynamic linkages among crop, biofuel, and energy prices using time series econometric techniques (e.g., vector-autoregression models or Granger causality tests) to determine if biofuels policies affected crop prices. Zilberman et al. (2013) and Serra and Zilberman (2013) provide excellent surveys of this literature. In sum, this literature argues that the extent to which oil and crop prices move together indicates the degree to which biofuels affect crop prices. One of the key findings, for example, is that there is no clear positive relationship between oil and corn prices, thus concluding that biofuels have had a minimal impact on corn (and other grain/oilseed) prices. The problem with the conclusion here is that time series analysis has a difficult time of separating out the effects of biofuels policies on crop prices; moreover, with a binding ethanol mandate, ceteris paribus, one would expect a negative relationship between oil and corn prices (de Gorter and Just 2009b).[11]

Therefore, this literature's hypothesis is mis-stated, a common error in most papers of this vast and burgeoning literature. Overall, this HFT econometric literature can best be summarized by recent surveys by those who at the same time continue to be major contributors to this literature. Zilberman et al. (2013, 278) conclude ' Biofuel prices do not seem to affect food commodity prices…but argue the quantity of crops used for biofuels does affect crop prices." Again, the impact of biofuels on commodity prices is through a simple shift in demand; price links in their conceptual framework play no role. But why did they conclude there was no link between biofuel and crop prices? Did they undertake the otherwise painstaking analysis and effort the discussion of Figure 7.1 prescribes. The answer is clearly not.

Serra and Zilberman (2013) continue to miss the point that if there is a link between crop prices and crude oil prices, biofuels (and biofuel policies) have their minimal impact; not maximal impact as the authors imply. Their tables 1 and 2 summarize 52 studies in the HFT econometric literature, with results that are all over the map.[12] The fatal flaw is revealed in the title of their tables: "Do biofuel or energy prices affect feedstock prices in the long run?" There is no distinction made between biofuel (e.g., ethanol) and energy (e.g., gasoline) prices. Yet, that is one of the key (but not only) takeaways from the analysis of Figure 7.1. One could expect a positive or negative relationship, depending on the regime (or type of market disequilibrium). Either way, one has to adopt our approach to tell which regime is binding and if there is market disequilibrium. However, the interpretation of the literature Serra and Zilberman (2013) are reviewing is entirely correct; to the last paper, they all assume that, if there is a strong positive price link, biofuels matter; otherwise not. And that is the exact opposite conclusion one should draw from any statistical analysis.

To emphasize the importance of the discussion above, consider a recent paper on the econometrics of the energy–biofuel crop price relationships by Kristoufek et al. (2012), who extend the literature by looking at non-linear relationships and conclude that the corn–ethanol (and biodiesel–rapeseed oil) price links did appear in 2007–2008 but returned to pre-food crisis levels afterwards. Mandate premiums kicked in after 2008, therefore, biofuels have an even bigger impact. Our analysis suggests the link remains very strong and any delinking can be explained by our discussion preceding Figure 7.1. Papers fail to use a lens or fail to use the correct lens to look through.

Most Studies Do the Same Thing

Unlike the econometric time series techniques reviewed in the previous chapter that use quarterly or annual data with economic variables as right-hand side regressors, the HFT literature is mostly deriving correlation/regression coefficients [to test for correlation using reduced form vector autoregression (VAR) models], which is plain old looking for simple time series relationships like correlations over time. All studies undertake stationarity tests and most studies find the prices are non-stationary. This inevitably leads to cointegration tests and vector error correction models (VECM) to capture possible non-stationary characteristics (in other words, the disentangling of short- and long-run dynamics in price interdependence or co-movement of prices in the long run).

Keeping with our corn–crude oil price example, some studies do not find the series cointegrated (Zhang et al. 2009; Serra et al. 2011; Dillon and Barrett 2014) whereas others do find cointegration. In fact, the results are all over the map not only for corn–crude oil but all other possible price pairings between each feedstock and either ethanol or biodiesel, or gasoline/diesel, or crude oil prices.

Cointegration is a reasonable way to look at this issue, but if there are mandate premiums and then a regime shift back and forth to a tax credit plus disequilibria in terms of ethanol-blending or ethanol-production constraints, then these tests says nothing, whether you find it or do not find it. For example, Zhang et al. (2010) use VECM and find sugar prices influence all other crops except rice, and that sugarcane ethanol influence short-run crop prices. This is really weak evidence; therefore, if sugar trends one way, others trend the same way (same time; no leading or no lagging—one simply cannot tell). The results can depend on the frequency of data (daily versus monthly).

A study by Myers et al. (2014, 992) find that co-movements between energy and agricultural feedstock prices tend to dissipate in the long run and which they claim have important implications for biofuel and food policy:

> Novel aspect of our approach is that we use common trend-common cycle decompositions to separate price movements into a component due to permanent changes in the underlying long-run equilibrium price for the commodity, and a component due to short-run transitory fluctuations around equilibrium.

Myers et al. (2014, 992) go further and add:

> We then develop and estimate measures of co-movement in each of these components across different energy and agricultural feedstock prices. The decomposition provides a different perspective on the extent of co-movement in energy and agricultural feedstock prices in both the long run and the short run—a perspective that cannot be obtained from cointegration analysis alone.

These tests are often augmented with Granger causality tests. Granger causality tests measure if one price moves, then another price moves. It finds out which price series moves first, but it could be akin to observing a batter in baseball flinching before getting hit by a pitch but flinching does not cause a batter to get hit by a pitch. Granger causality tests to not take into account the looking phenomenon, which happens all of the time in commodity markets.

The finding by Zhang et al. (2010) (discussed in Chapter 6) for a lack of short- or long-run causality between fuel and crop prices except sugar prices cause crude oil prices could simply mean it is faster for ethanol production to react than crude oil production (distribution), assuming the tax credit is binding (which implies crude oil causes ethanol prices; not the other way around, as is suggested).

However, again, the many studies employing Granger causality tests do so with no regard for regime changes: we showed corn prices cause ethanol when the mandate is binding and vice versa if the tax credit is binding. No regard for this is taken into account in the literature. Meanwhile, Balcombe and Rapsomanikis (2008) find that crude oil prices determine the long-run equilibrium prices of both sugar and ethanol where sugar prices Granger-caused ethanol prices (this evidence is consistent with the blend mandate binding in Brazil).[13] No wonder the results are inconclusive or varied between studies with no consensus being able to be reached.

Other studies test for structural breaks, using, for example, smooth shifting auto regressions, and many find structural changes in the 2005–2008 time period (e.g., Enders and Holt 2012 find that corn prices had a structural change in August 2006, coinciding with our September 2006 date, particularly if one evaluates weekly data in Figure 4.1 where corn prices were at their lowest in August). de Nicola et al. (2014, 11) find an additional structural break in May of 2007 and that energy and agricultural prices are positively correlated, reflecting:

> The fact that related commodities are generally close substitutes in demand, share common market information and have similar input costs. Yet, we find an overall increase in the degree of co-movement between energy and agricultural price returns in recent years, especially in the cases of maize and soybean oil.

Wetzstein and Wetzstein (2011) argue the strong connection between crude oil and agricultural prices is a "myth" [which we would agree with when mandates are binding and biofuel policies have their strongest impact, but that is not what Wetzstein and Wetzstein (2011) are arguing]. They argue that biofuel production and, therefore, feedstock demand are less responsive to energy price changes because of capacity constraints. We emphasize the same phenomenon in this book but that does not nullify biofuels policies impact on crop prices or our framework; on the contrary, any kind of capacity constraint emphasizes the importance of biofuel policies and their roles in affecting prices. Indeed, one has to adopt our framework to truly tell when capacity constraints occur in production, marketing, and blending of biofuels (as the discussion of prediction errors in Figure 2.1 of Chapter 2 emphasizes).

Shortcomings of the HFT Literature

Time series models do not impose any theoretical structure and mainly focus on empirical investigation of price links. This HFT econometrics literature is inconclusive about biofuel policies' role because it fails to recognize that the crop–biofuel link began in September 2006 (the time series analysis in this literature typically ends in 2007 or 2008) and has not controlled for the switch in regimes between a tax credit/tax exemption and a mandate or when US biofuel prices are determined on world markets through exports or imports (or both). There is no account of the three types of capacity constraints their statistical analysis: the ethanol production- and marketing-capacity constraints (delinking corn from ethanol prices and falling below ethanol prices) and ethanol-blending capacity constraints (delinking ethanol from gasoline prices and falling below gasoline prices on a mile per gallon basis). The results differ across studies for various reasons: different time periods are used, as are different countries, data frequency, and different combinations of prices depicted in Figure 7.1 for each of biodiesel and ethanol prices, including a trend or not, suppressing the constant or not, and using nominal versus real prices.

The analysis done by this HFT literature is only an indirect test of biofuels impacts on crop prices. Regardless, this approach would provide little economic insight as to how specific policies impacted prices because it has so far been unable to control for the switch in regimes between a tax credit/tax exemption and a mandate. We showed that ethanol (and, thus, corn) prices are negatively related to gasoline (and hence crude oil) prices when the mandate is binding (positively related otherwise). Equation (2.1) in Chapter 2 specified a very specific relationship (multiplier between corn and ethanol prices that varied with the costs of production of ethanol and the relative price of dried distillers grains with solubles (DDGS) to corn (which varies with respect to wheat prices and other prices of higher protein crops). As Figure 2.2 shows, by August 2007, corn prices were finally linked to ethanol prices. From that point forward, corn (and oilseed) prices were more closely related to their biofuel price counterparts. As a result, any time series analysis that begins well before 2006 and ends in either 2007 or 2008 misses some key episodes in the process of biofuels policies linking corn (as well as oilseed) prices to energy prices. It should also be noted that ethanol prices can float up and away from gasoline (crude oil) prices when mandates are binding for periods of time, delinking corn and oil prices.

Harri et al. (2009) use futures price data while most studies use spot or cash data for prices. Some studies are analyzing the short run whereas others are analyzing the long run. There is also no test to differentiate between the biodiesel–soybean and sugarcane–ethanol models from corn–ethanol. In later chapters, we present these models and as the readers will see for themselves these are fundamentally different from each other and from corn–ethanol production. For corn–ethanol, the co-product DDGS reinforces the corn price increase due to ethanol policy. However, soybean meal is a joint product with soybean oil used in biodiesel production. As soybean oil and soybean prices increase, the production of soybean meal increases too, but its price declines, thereby moderating the soybean price increase due to biodiesel policy. Sugarcane–ethanol revolves around competing products (sugar and ethanol) produced from non-traded sugarcane with two by-products: molasses from sugar production used as ethanol; and bagasse used for electricity production. The price transmission of a crude oil price shock is, therefore, even more muted on sugar prices than it is on soybean prices. The point is the HFT literature does statistical tests on all three types of biofuels yet does not recognize the very different economic structures of the three sources of biofuels.[14]

In a recent book entitled *Flash Boys*, Lewis (2014, 209) describes the (apparent) continued shenanigans in modern finance markets which prompted one of the protagonists, a high frequency trader on Wall Street, to suddenly realize (shortly after Wall Street switched from floor trading to electronic trading) that:

> All of a sudden the market is all about algos and routers. It's hard to figure this stuff out.

Lewis (2014, 93) describes how the perplexed Wall Street high frequency trader noted that:

> The role had been spawned by the widespread belief that traders didn't know how to talk to computer geeks and that computer geeks did not respond rationally to big, hairy traders hollering at them.

The HFT papers surveyed in this chapter are very statistical method intensive exercises, in other words, mostly about "algos and routers" where "computer geeks...[do] not respond rationally to big, hairy traders hollering at them." We take on the role of the latter in this chapter. Therefore, given all of these factors, it should be of no surprise that overall, the econometric approach on the role of biofuel and energy prices are inconclusive.

Concluding Remarks

In general, econometric studies have problems identifying the effect (in other words, a positive input channel effect versus a negative effect due to the mandate—therefore, an overall net positive, but cannot decompose the effects). If there is no change in the mandate over time (or the variation is small), then one cannot identify the effects of biofuel policies. Only when the tax credit is binding could the HFT methods identify the effects of biofuel policies. It is also difficult using econometric methods to identify (short) periods when individual policies are binding.

But in their abstract, Enders and Holt (2014) note:

> [T]he introduction of ethanol as an important fuel source has likely contributed to the run-up in grain prices.

Therefore, all is not lost in this literature; at least one study concluded that ethanol "likely contributed" to the corn price increase; this is very reassuring. Corn prices are locked onto crude oil prices through the tax credit (this is the lowest corn prices can go unless there are ethanol production, marketing or blending capacity constraints); or corn prices float up and away when mandates are binding (or if world prices dictate domestic prices through trade). This latter case has a negative relationship between crude oil and corn prices. However, this ignores the input cost channel, therefore, the net effect is probably net positive. Indeed, all of the possible channels we discussed in Figure 7.1 have to be evaluated for each of the two regimes; this is the implication of our analysis.

This distinction is very important for several reasons. First, the causality between corn and ethanol is regime-dependent. Second, the impact of supply/demand shocks on ethanol and hence corn prices are also regime-dependent. If ethanol prices are locked onto gasoline prices through the tax credit, then supply/demand shocks will have negligible price effects—only changes in oil prices can affect corn prices. Finally, the stronger the positive relationship between prices, the *smaller* is the impact of biofuels on food commodity prices. This is the opposite hypothesis posed by the HFT econometric literature. In fact, this entire literature has to start all over again and carefully assess all of the economic pathways we defined and discussed in Figure 7.1 before any conclusions can be reached, let alone deriving policy implications therefrom. Furthermore, this includes analysis of various simultaneous effects and also of possible market constraints that can break the price relationship one thinks is supposed to exist. Moroever, the only way to find out properly when there are market constraints (like an ethanol-production-capacity constraint) is to follow the approach developed in this book.

CHAPTER 8

THE IMPACT OF BIOFUEL POLICIES ON FOOD COMMODITY PRICE VOLATILITY

The controversy over biofuel policies has sparked debate not only on what caused the high levels of grain and oilseed prices, but also the alleged increase in price volatility and its potential deleterious effects. To illustrate, take two headlines, each involving a Cornell professor:

> "Cornell's Per Pinstrup-Andersen: Don't Believe the Hype (and Data) Surrounding Food Price Crises. Cites Food Price Volatility as Greater Danger than High Food Prices," an International Food Policy Research Institute (IFPRI) news headline on Per Pinstrup-Andersen (2013b)

Whereas, the title of a paper in *Foreign Affairs* by Barrett and Bellemare (2011) reads:

> "Why Food Price Volatility Doesn't Matter: Policymakers Should Focus on Bringing Costs Down"

The passion over volatility has been intense since the 2007–2008 price boom, as most international organizations, like the Food and Agriculture Organization of the United Nations (2008a, 2009), High Level Panel of Experts on Food Security and Nutrition (HLPE) (2011), and the Interagency Report of the G-20 have put volatile crop prices high on the agenda and have identified this issue as a policy priority. These organizations have emphasized the detrimental effects of price volatility and adopted the explicit goal of exploring its causes and consequences. There has been a flurry of activity among governments and international organizations to do something about it, and various initiatives have been undertaken. The Interagency Report to the G-20 (2011) argues for improving the functioning of global commodity markets as a policy priority and calls for a global perspective in policymaking. The World Bank called for the international aid community to coordinate their activities in order to compile useful and real-time information for policymakers when examining the availability of key resources to ensure adequate food for those most affected.[1] The increasing concern over food price volatility hinges on its impact on food security as it adversely affects household incomes and purchasing power of poor consumers, potentially transforming vulnerable people into poverty and hunger. Meanwhile, farmers in the hinterland are alleged not to be able to take advantage of high prices in market centers.

Barrett and Bellemare (2011) chronicle worries by national governments over price volatility like that of French President Nicolas Sarkozy, who warned that "if we don't do anything, we run the risk of food riots in the poorest countries" and of think-tanks like the Brookings Institution with an opinion piece stating that "the crux of the food price challenge is about price volatility rather than high prices per se" and that it is "the rapid and unpredictable changes in food prices that wreak havoc on markets, politics and social stability" and that "In late June, the G-20 released a ministerial declaration calling on members to end food price volatility."

"Ending" price volatility is a difficult thing, especially when there is even a controversy as to how to define it. Some argue it means volatility around the mean, which is unobservable and needs to be estimated. However, such a definition raises a number of questions, most important of which is how should we actually measure volatility? Should the absolute or relative dispersion be used, and if so, which mean to use: expected of unexpected? Or should we, perhaps, simply focus on absolute or relative changes in commodity prices as indicators of price volatility?

The literature generally understands volatility as unexpected price changes and analyzes it by modeling the commodity price levels and spillovers between them, using the Generalized AutoRegressive Conditional Heteroskedasticity (GARCH) model as the main working tool. As noted earlier, volatility is difficult to define and measure and is subjective. For example, if one defines volatility in absolute terms, then clearly it is bound to rise when prices are much higher. On the other hand, if a researcher defines volatility relative to the mean, he or she immediately faces a problem of choosing the "correct" one. It could be the mean based on recent data; the mean to be reverted to; or even the time trend in the commodity prices. In addition to the selection of the mean, one also faces a problem of treating the prices for the analysis: should real or nominal prices be used? And should they be detrended? Other important issues related to price volatility include the selection of the time horizon and the data frequency; accounting for cycles versus trends; and distinguishing between volatility and uncertainty. With respect to the last point, if the market is subject to mean reversion, then prices are more predictable but we may not be in a "mean reversion" world now, unlike up to 2005 when real prices were continually declining (Figure 1.3).

The quotes from the beginning of the chapter on two different Cornell professors reflect the tension over price volatility in the profession. There has been a good deal of research on the link between food prices and political instability (e.g., Barrett and Bellemare 2011; Bellemare 2015; Barrett 2013; and Berazneva and Lee 2013). Based on their own research and that of others, Barrett and Bellemare (2011) conclude that it is the level, not volatility of prices, that causes political instability. In fact, poor households can benefit from price instability, depending on the extent to which they can switch to other products whose price has not risen. Like the commodity coverage in this book, Bellamare et al. (2015, 882) analyze only a subset of a "potentially infinite set of commodities consumed and produced by the household"—although they do include *teff*[2] (a commodity we sadly ignore). Because we show grain and oilseed prices (which constitutes the bulk of calories consumed by mankind either directly or indirectly) move closely together, this would not allow poor consumers to benefit much from arbitrage across commodities as prices rise.

Price volatility has also been the focus of many economic studies by international organizations like the Organization for Economic Co-operation and Development (e.g., Huchet-Bourdon 2011) and the International Centre for Trade and Sustainable Development (ICTSD) (e.g., Tangermann 2011), and also of a host of academic studies with "volatility" in their titles (like this book!)[3] Others believe excessive price volatility is just a perceived problem whereas others still argue against the view that price volatility in food commodities has risen systematically in the biofuel era (e.g., Baumeister and Kilian 2014; Balcombe 2011; Gilbert and Morgan 2010).

The primary economic arguments against more price volatility is that it makes it difficult for farmers (particularly in developing countries) to plan, and may lead to dire circumstances when unexpectedly low prices prevail. However, when we look at Figure 1.3, we see that all food grain price shocks historically are above (and not below) the long-term negative price trend for food grains. On the face of it, this might suggest that farmers, in fact, benefit from price volatility as it tends toward giving farmers unexpected windfalls rather than unexpected shortages of cash.[4]

Consumers face the flipside of price volatility. When prices spike upward, consumers can only achieve the same level of consumption by trading off other goods. However, if all food commodity prices move together (as we saw in Figure 1.2), consumers that spend large portions of their income on food (as more than one billion of the world's poor do) are unable to take advantage of volatility and so undernourishment may result. The long-run rise in prices due to biofuels could have substantial impacts on the world's poor that are not themselves involved in agricultural production.

The issue of price levels versus volatility is an age-old debate that involves the issues just identified and many more, none of which really concerns us here. All we want to do is assess what implications biofuel policies have for food commodity price volatility. To do so, we want to focus on the implications of the two states of nature and the source of shocks: corn supply and oil prices. The literature on volatility and price levels has focused almost entirely on the traditional sources of volatility like commodity specific supply and demand shocks, developing country policy responses, preexisting farm and trade policies, changes in macroeconomic variables spilling over into commodity markets, the financialization of commodity markets, and the interaction between these shocks and commodity stock levels for intertemporal price smoothing (Carter et al. 2013; Wright 2011, 2014; Bobenrieth et al. 2014; Janzen et al. 2014).[5]

In prior chapters, we showed that these factors can, in fact, affect prices and, therefore, price volatility. However, these factors can only lead to pushing prices up by the amount of the mandate premium and only when the mandate is binding. The mandate premium was zero in the year 2008 during the price run-up and the latter run-down. This leads us to believe that corn prices at that time were lowest when locked onto crude oil prices. In this case, these traditional pathways for price turmoil would be completely impotent.

Biofuel policy has ushered in a new economics of volatility. By linking biofuels to crop prices, the economics of commodity price volatility have become more complex. The interaction effects now not only depend on the source of the shock (oil prices versus crop supply and demand shocks) but also on which biofuel policy determines the world biofuel market price (e.g., tax credits versus mandates). In addition, world prices and trade can impact which domestic policies are binding and the extent of any mandate premium over a tax credit. The domestic mandate is still important in that it takes a required amount of feedstock for biofuels.

We first analyze the implications of our theory for food commodity volatility, comparing food commodity supply shocks with crude oil prices shocks under the two alternative biofuel policy regimes: the tax credit and the mandate. After a brief glimpse of the high-frequency time (HFT) series literature on volatility, we devote an entire section on the relevance of the competitive theory of storage—the proposition that quantity shocks explain price levels and volatility through the demand for storage theory. We show that, although the price rise due to biofuels impacted the storage decision, modeling this shift in the demand for storage requires explicit recognition of the theory of biofuel policy as we have developed it so far in this book.

Biofuel Policies, Market Shocks, and Grain-Price Volatility

It follows naturally form our theory that the volatility of corn prices, caused by corn-supply or oil-price shocks, critically depends on what policy is binding: a tax credit or a mandate. When a fuel blender's tax credit is binding, the corn price is directly linked to the oil price. In addition, as we described earlier, for every $0.01 per gallon increase in ethanol prices, corn prices increase almost $0.04 per bushel. For such a binding tax credit, exogenous shocks to corn supply or demand (such as bad weather, exchange rate depreciation, increasing demand for corn in developing countries, or the initial levels of stocks in the supply-of-storage model) have comparatively smaller effects on the corn price level or on corn-price volatility than the behavior of oil prices (de Gorter and Drabik 2012a). However, the volatility in the oil price gets translated into corn price volatility, because oil prices determine ethanol prices (through the binding tax credit), and ethanol prices, in turn—under any biofuel policy—influence corn prices (Drabik 2011).

However, if a biofuel blend mandate determines the ethanol market price,[6] corn prices respond to corn supply and demand shocks. This happens because the blend mandate significantly weakens the link between ethanol and oil prices. In fact, higher oil prices mean lower corn prices, because fuel consumption declines. Given a fixed ethanol blend requirement, ethanol production contracts as well, with the effect of lowering both ethanol and corn prices. However, a countervailing effect also exists: a higher oil price translates into higher energy costs in corn production and results in higher corn prices (Baffes and Haniotis 2010). The net change in corn prices due to a crude oil price increase is therefore indeterminate a priori.

Yano et al. (2010, 2011) model volatility using the blend mandate model and determine that a corn supply shock has a much larger impact on corn prices under a mandate than under a tax credit (and vice versa for a shock in gasoline prices). With a corn supply shock, the variability of prices increases 1.6-fold compared to a tax credit.[7] Hertel and Beckman (2012) employ a consumption mandate model and get almost identical results (more due to the different design of the experiment and baselines than due to the similarity of models used) whereas McPhail and Babcock (2012) find mandates also cause more price volatility, but with a much lower impact than the other two studies cited. Both Hertel and Beckman (2012) and McPhail and Babcock (2012) present results on the implications of the blend wall, but model it incorrectly as in Tyner and Viteri (2011) and NAS (2011) (the reader will have to wait until Chapter 10 on how to model the blend wall correctly).

There is a burgeoning literature on food grain/oilseed price volatility using HFT econometrics, but do not take into considerations just discussed about the importance of regimes, price linkages and the nature of any disequilibriums. The most common approach is to use a parametric volatility model, commonly called the GARCH model, which estimates volatility using historical data and focuses on market interactions in terms of conditional second moments. Although some papers argue that crude oil, gasoline, and ethanol market shocks do not spillover into grain prices (e.g., Qiu et al. 2012; Zhang et al. 2009), others do find spillover effects (e.g., Serra 2011; Serra et al. 2011a, b; Gardebroek et al. 2013; Gardebroek and Hernandez 2013; de Nicola et al. 2014).

The Implications of the Theory of Competitive Storage

The theory of competitive storage emphasizes the nonlinear relationship between prices and stock levels. The lower the stock levels, the higher the sensitivity of prices to "normal" supply/demand shocks. Higher price levels are therefore associated with higher price volatility, and, therefore, are important for the topic of this chapter. The theory rests on the notion that if you can store grain until a future date, you will only sell if expected future prices will not, at the very least, cover the cost of storage. Thus, speculators increase stocks when negative price shocks have led to prices that are below expectations. Alternatively, one will sell their stored commodities when prices are higher than future expected levels. The level of stocks held by firms in a rational expectations competitive market are such that the current price equals the discounted value of the future expected price. This stabilizes the price for any commodity supply or demand shock. The market equilibrium is characterized by a downward-sloping demand for storage (or inventory) curve that is added horizontally to a very inelastic grain or oilseed consumption demand curve. Total demand is much more elastic than demand for consumption because of the demand for storage, moderating the price changes from supply or demand shocks when stock levels are adequate or high ("normal").

Under the theory of storage demand, we could expect one of two responses to increases in commodity prices. If the increases are believed to be due to a temporary shock, we would expect stocks to be sold and a depletion of storage because speculators would anticipate a return to normal prices. If, instead, the increases are believed to be due not to a temporary shock, but due to a long-run change in the price of the commodity (say, due to an unexpected change in demand due to biofuel polices), then we would expect the opposite to occur. If speculators expected that prices would continue to rise, they may retain their storage, waiting for better prices in the future. When long-run price expectations increase, the demand for storage at each individual realized price increases whenever positive.

The first study by Timmer (2009, 16) emphasized the importance of the supply of storage model in explaining price movements in the modern era when an unexpected permanent shock occurs: "as rice prices increase, so do stocks…consistent with modern supply of storage theories." Since then, further applications have been developed in Wright (2011, 2014), Hochman et al. (2011), Carter et al. (2011, 2013), Bobenrieth et al. (2014), and Janzen et al. (2014) although many qualitative studies emphasize low stocks means higher price sensitivity to shocks (Mitchell 2008; Abbott et al. 2008, 2009).[8]

The basic idea is that an unexpected permanent shift in demand for a crop will affect the level and volatility of prices differently in the short run versus long run, depending on the timing of the realization when the shock is permanent, initial stock levels, and market parameters. Thus, there is a transition, and the timing varies, depending on the circumstances. Carter et al. (2013) also argue demand for inventories increased in anticipation of the ethanol demand, causing baseline prices to be higher than otherwise,[9] and buffering the price increase in 2009–2010.[10] Wright (2011, 2014) and Bobenrieth (2014) do not give any empirical estimate of the impact of the biofuel demand shock on corn prices other than to say that their model can cause "explosive" price runs ("bubbles" or "spikes"), depending on the parameters of the model. Carter et al. (2013) generate the highest corn price increase of 30 percent due to storage behavior, whereas Hochman et al. (2011) get lower price impacts and Roberts and Tran (2012) get a 23-percent price increase.[11] Janzen et al. (2014) find that wheat prices were most significantly affected by supply/demand shocks, and stock build up was the second biggest contributor.

There has been some confusion as to when stocks actually increased. Figure 8.1 shows that total world stocks declined for two years (2005–2006 and 2006–2007) and only increased in 2007–2008 (along with prices, the only year prices and stocks went in the same direction).[12]

Prices began to rise in September 2006, as we have shown. However, biofuels and biofuel policy's impact in creating a crop–biofuel price link (with huge multipliers) was the only true surprise to the market. China's gross domestic product had grown over 9 percent per year since 1978 and continued growth was already built in to future price expectations. Biofuels, on the other hand, can be labeled in its initial impacts as an unexpected shock as no one had anticipated at which crude oil prices biofuels (and particularly ethanol) would suddenly sweep all grain and oilseed prices into lock step. Moreover, it took the market substantial time to determine that the impact of biofuels on prices was not going away. In the intervening

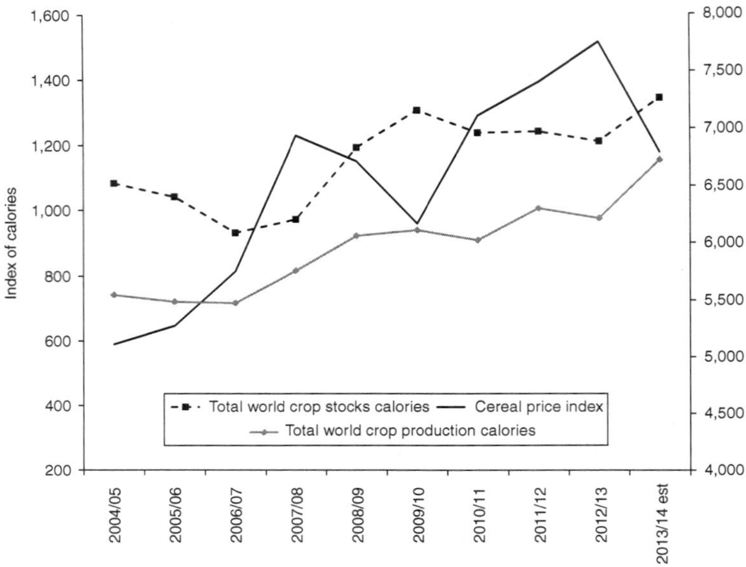

Figure 8.1 Cereal prices versus grain and oilseed stocks and production (in calories).

Source: WASDE (various issues), World Bank (2014).

period, speculators behaved as if the price rises were temporary. Therefore, inventories were depleted. Some blame the reduced storage for the apparent increase in price volatility, arguing that relatively routine supply and demand shocks now appeared to have caused huge changes in prices, compared to normal times. This is the argument made by the literature on the theory of competitive storage.

This argument has validity, but only when the grain and oilseed prices are *not* locked onto crude oil prices. However, prices were locked onto crude oil prices in 2008 when the demand curve for a grain and oilseed was essentially flat. The only slope to a demand curve for these commodities would represent the effect of a *change* in biofuel consumption on crude oil prices—a modest effect at best. Janzen et al. (2014) claim wheat price increases in 2008 were mostly due to sector specific net supply shocks of wheat sector, the impact of which worked through stocks. We have outlined previously how grains and oilseeds are very substitutable, and, thus, there is a law of one relative international price of grains and oilseeds such that a net supply shock in one sector does not remain isolated in that sector. Net supply shocks for wheat in 2008 and stock levels had little to do with wheat prices—prices would have gone up regardless due to the substitution with corn or other grains. In our framework, the theory of stockholding only has legitimacy to the extent that the mandate premium over the tax credit is positive.

To illustrate the impact of storage, Byrd (2012) developed a simulation that incorporates both the theory of biofuels and the theory of competitive storage. The latter is generally only applied in simulations due to the tremendous computational demands of the theory. In order to find an analytical solution for such a theory requires limiting the model to just a few periods of time, and placing some severe limits on the potential prices that could be realized. If you employ more realistic assumptions, the model quickly becomes too unwieldy to solve analytically, forcing us to rely on sophisticated numerical algorithms to solve the model and create predictions.

Byrd (2012) builds her model by mimicking the procedures outlined in Williams and Wright (1991), but augmenting the price formation theory to incorporate the biofuel market model we have outlined in previous chapters. These price links—and in particular the points at which price links break or are made—have important implications for the formation of price expectations. In particular, the distribution of prices above the oil price will be determined by the standard shocks to grain and oilseed markets. Below this level, the oil market determines the distribution of the price. The model employs all three regimes based upon the US markets for corn, ethanol, and crude oil.

1. If no ethanol is produced, then the realized price of corn is determined by the intersection of corn supply and the sum of the consumption demand and storage demand curves. If the model is parameterized such that ethanol is never produced—no oil link is ever formed and our storage model is identical to that of Wright and Williams (1991).
2. If ethanol is produced and there is either no mandated blend or the mandated blend is not binding, then the tax credit is binding and price of corn is determined indirectly by the price of gasoline. We assume the price of ethanol is determined so that price per mile traveled using ethanol is identical to that of gasoline, and the price of corn so that ethanol blenders achieve zero profits, as presented earlier in the book.
3. If a mandate is in place and binds, then the price of corn is determined jointly by the price of blended fuel and the amount of corn needed to meet the mandate. The price is determined where the supply of corn is equal to the demand for consumption plus the demand for storage plus the amount required to satisfy the mandate for ethanol and so that fuel markets clear.

Practically, the maximum of the three possible prices resulting from these three regimes is the price that will prevail. Constant elasticity curves for the supply of corn, the supply of gasoline, the demand for miles, and the demand for corn consumption are assumed. The model is calibrated for the 2010–2011 crop year.[13] Moreover, the oil price and the amount of corn produced vary randomly (mimicking actual empirical distributions).

Figure 8.2 displays the simulation for the year 2010–2011 with and without a blend mandate. As you can see, the presence of the mandate does increase the storage. Recall that under the mandate, the amount of ethanol demanded is jointly determined with the fuel supply and demand for miles, and that the price of corn will exceed the price suggested by the gasoline equivalent depicted on the figure when the mandate is binding. In fact, there is substantial storage even above corn prices of \$8 per bushel. Additionally, the total corn demand (including storage) is indeed more elastic than the simple corn consumption demand. However, the mandate does decrease this elasticity (the total demand has roughly the same slope but much larger

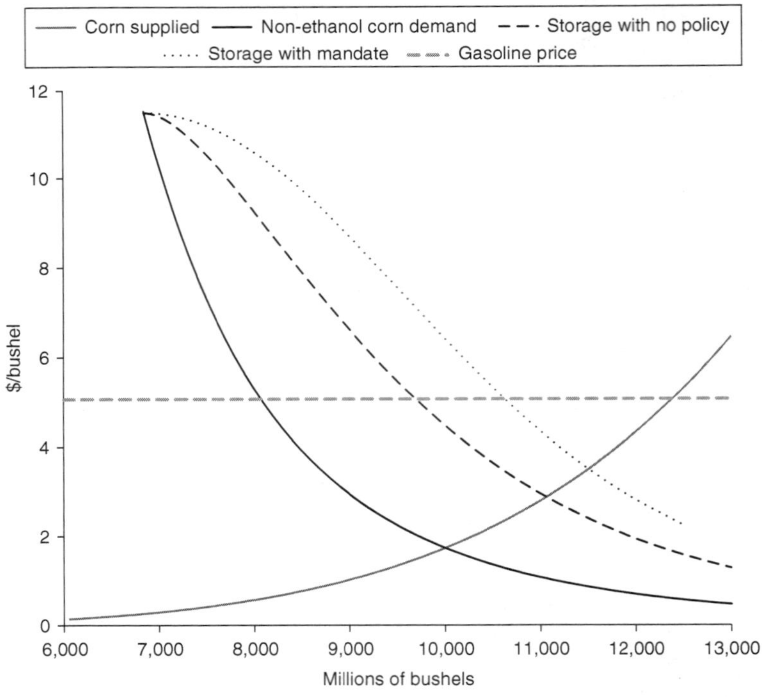

Figure 8.2 Demand for storage 2010–2011 crop year.

Source: Byrd (2012).

quantities). Because the mandate implies roughly the same change in prices for a given change in quantity traditional supply shocks would have roughly the same impact on markets with and without the mandate.

Figure 8.3 displays the same simulation, only comparing storage demand with and without the tax credit in addition to the mandate. The additional storage for the tax credit in addition to the mandate is much more minor, but the results are roughly identical. The amount of storage increases at each price level, but the elasticity declines over most relevant prices. Moreover, the impacts of supply shocks do not appear to be affected in most relevant ranges. This is suggestive that, to the extent we saw an increase in volatility of prices, it was not due to the storage effect of the biofuel policy on price responsiveness, but due to the price links induced by the policies themselves. It is the connection to volatile oil markets and the high multipliers

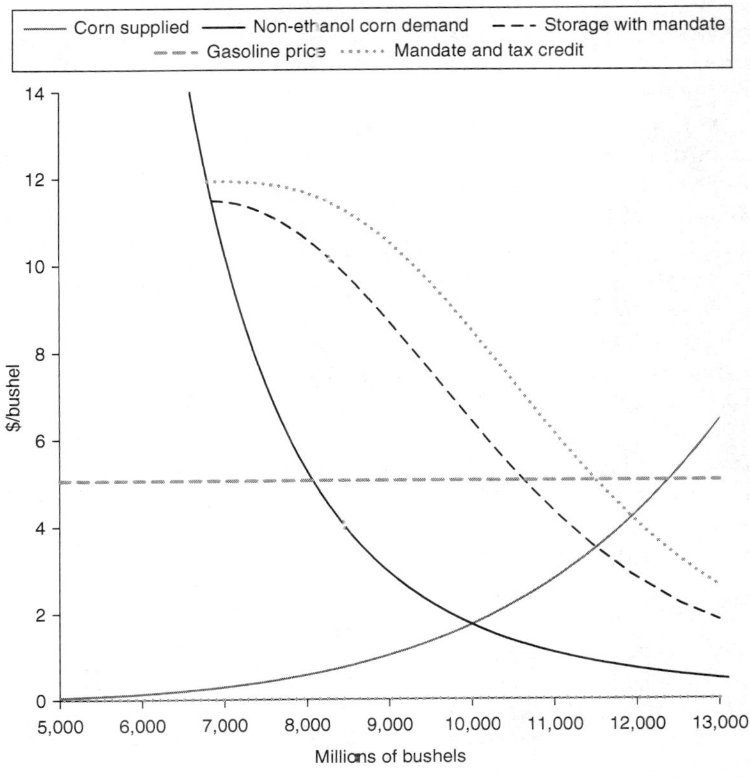

Figure 8.3 Demand for storage under a mandate with and without a tax credit.

Source: Byrd (2012).

that are driving the price volatility—not a continuous storm of relatively routine supply shocks.

Our results on the supply of storage differ from those specified by the literature (studies mentioned above) that ignores biofuel policy regimes and price linkages. In essence, crude oil places a (somewhat volatile) floor under the corn price through the tax credit and the mandate. Without recognizing these links, there is no such floor, and the storage function takes on a much different shape which implies more and more volatile prices as the demand shifts out. However, much more research is required on this important model of biofuel policies and the supply of storage.

Evolutionary Assessment of Biofuel Policies and Grain-Price Fluctuations

It is important to recognize that the first two categories of biofuel policies, tax credits and mandates, do not, by themselves, discriminate against international trade. However, the other policies do, and the fact that they were in place for over 30 years in the United States explains, in part, why we have observed increasing corn–ethanol production in the United States (rather than, for example, import of less expensive sugarcane-ethanol from Brazil). Such anti-international trade policies also partly caused US ethanol production to increase sharply after 2005, with the result of not only exacerbating the increase in world food grain commodity prices but also widening the gap in distribution of welfare gains and losses between developed and developing countries.

Imported sugarcane-ethanol from Brazil was unable to supply the ethanol market in the near term (although it was considerably less expensive than domestic corn-ethanol). Brazil needed time and significant investment to increase sugarcane production and expand infrastructure to logistically move ethanol to ports. Meanwhile, US corn-ethanol production increased sharply, as farmers simply converted land from other uses to increase supply immediately. The lesson here is that long-term policies have short-term implications. If no trade-discriminatory biofuel policies had existed, more investments in foreign ethanol production would likely have taken place. In Brazil, for example, the sugarcane crop would almost certainly have had a higher production base, and more ethanol infrastructure would have been built to supply world sugar and ethanol markets. Likewise, if US environmental policies had been implemented more gradually, US biofuels would not have had such a dramatic effect on commodity prices. In this respect, US biofuel policies have continued the saga of distortionary agricultural policies that not only cause world price instability but also have harmed the consumer welfare of food-importing developing countries.

US agricultural and macroeconomic policies are relevant, but so are environmental and energy policies. All of these policies have strongly affected food grain commodity prices and volatility. The roller-coaster ride in market volatility since the financial crisis of 2008, and the re-emergence of peaks in corn, wheat, and soybean prices in 2011, can be traced to all three forms of intervention: US renewable-energy, environmental, and agricultural policies. There can be little doubt that corn prices would not have reached a low point of $3.30 per bushel in 2009–2010 if US renewable energy and biofuel policies had not influenced the market. This lower bound contrasts sharply with the low of $1.62 per bushel for corn in 2005–2006, when the global economy was enjoying an unprecedented economic boom.

Along with the nexus between the impacts of historical US agricultural and biofuel policies on program commodities (especially corn, soybeans, and wheat), the heavy-handed US governmental intervention and global sugar systems also influenced world corn prices. The US import quotas and internally insulated sugar price supports initially depressed world sugar prices, indirectly subsidizing Brazilian sugar-ethanol production, which, in turn, responded to increases in crude oil prices since the mid-1970s. The destabilized world sugar market that resulted had chilling effects on worldwide production of sugar and on sugar prices and helped lead Brazil to reallocate sugarcane production away from ethanol to sugar markets. Currently, US biofuel policies are helping to stabilize Brazil's ethanol market, which, in a dynamic legacy perspective, was destabilized by historical US sugar policy.

Concluding Remarks

Serra and Zilberman (2013, 142) argue "In spite of the progress made in theoretical modeling of food–energy price relationships, there is no widely accepted model that explains food price volatility." We would agree with that assessment but believe two models have to be front row and center in any analysis of food commodity price volatility: the theory of biofuel policy as we have presented it so far in this book (the various price links and the blend mandate model) and the theory of competitive storage where the total demand is highly nonlinear as stocks decline, the levels of the latter of which were depleted unexpectedly because of biofuel policy shock to the market. We cannot make any conclusion as to how biofuel policies have impacted price volatility; the only thing we have noticed with biofuel policies is that change has been the only constant. The price links should be the focus of attention and the specific complexities of the myriad of biofuel policy instrumentalities employed and a plethora of interaction effects between biofuel policies themselves within or across countries, between biofuel policies and other policies (in particular, farm subsidies and developing country policy responses to the high prices), and between biofuels and between commodities.

The instability from interaction effects between polices within a country and across countries will be illustrated in the penultimate chapter of this book that an expired US tax credit with exports dragged down Brazilian ethanol prices in early 2012. We also show in Chapter 9 on biodiesel how the US "splash and dash" program created price volatility, including the effect of the EU import tariff in response. In Chapter 10, we will show how the complex nested mandates in the United States and the ethanol blend wall has caused instability, especially with the Environmental Protection Agency clawing back the renewable fuel standard in November 2013 for 2014 and the ensuing uncertainty if they will change their mind (to be decided soon at the time of this writing). We showed how two types of capacity constraints: the ethanol production-capacity constraints and ethanol blending-capacity constraints, the latter causing ethanol prices to fall below gasoline prices on a miles per gallon basis for three short time periods in the past. We showed this had implications for price volatility. For biodiesel, we will show our model over predicted biodiesel price in 2011 because of sugarcane price spike in Brazil and, thus, a biodiesel production constraint (because biodiesel and sugarcane compete in the advanced sub-mandate)—and again in 2013 because of all renewable identification number prices equalized and biodiesel helping overcome blend wall. We also showed in Chapter 2 that we had several prediction errors for ethanol because of various reasons, causing production constraints, thereby affecting price volatility.

Meanwhile, US and EU import quotas for sugar destabilize world sugar markets, causing Brazil to divert sugarcane from ethanol to sugar, thereby destabilizing ethanol markets. Therefore, a complete analysis of how biofuel policy affects commodity market price volatility is still forthcoming.

CHAPTER 9

THE ECONOMICS OF BIODIESEL AND THE CENTRAL ROLE OF THE EUROPEAN UNION'S POLICIES

Biodiesel averaged 20 percent of total world biofuel production by volume in the period 2010–2012.[1] Biodiesel is expected to reach 25 percent of total biofuel volume by 2025. However, its share in vehicle miles traveled is higher because a gallon of biodiesel obtains 91.3 percent of the miles traveled compared to a gallon of diesel whereas ethanol produces an estimated 70 percent of the miles traveled as gasoline—and diesel engines, on average, get approximately 37 percent more miles per gallon than gasoline (see Box 2.2). The European Union, United States, and Brazil are, respectively, the world's largest producers and consumers of biodiesel. The European Union consumed 42 percent of total biodiesel in 2013 (down from 75 percent in 2005), and it is expected to decline to 40 percent in 2020 (OECD/FAO 2014).

As we noted in Box 1.2, the European Union does not have a biodiesel mandate per se; it has an overall target that can be filled with any combination of biodiesel or ethanol (and to date has not been enforced in every country). The European Union has been wavering on its mandate. In July 2013, the European Parliament's Environment Committee voted in favor of a 5.5 percent cap on the target for first-generation biofuels[2] (significantly below its first target of 10 percent) whereas Spain at one point reduced their target of 7 percent to 4.1 percent.[3, 4] The United States has a specific biodiesel mandate that is nested in the overall mandate, the (complicated) details (and impacts) of which will be explained in Chapter 10. Brazil's biodiesel mandate of 5 percent is filled by domestic production, mostly from soybean oil as the complicated government bidding system generally precludes any imports.

In the United States (European Union), biodiesel prices have been on average 1.95 (1.82) times higher than ethanol prices in the period from April 2007 to December 2012 (Table 9.1).[5] A similar ratio is found in Brazil.

There are several possible reasons for this significant price differential. There have been partial import bans on Asian exports of biodiesel to the European Union; the United States bans biodiesel from palm oil because the Environmental Protection Agency (EPA) determines palm oil not to meet the required binary sustainability threshold. The US and EU policy requires biodiesel to reduce greenhouse gas (GHG) emissions by 20 percent and 35 percent, respectively, relative to the diesel it is assumed to replace. The European Union will allow palm oil from specific plants in Asia that meets the 35-percent threshold; the United States has a blanket or global policy on palm oil and, therefore, none is used for biodiesel, although the US threshold is lower than that of the European Union. Ironically, US corn-ethanol has not satisfied the EU requirement that it reduce GHG emissions by 35 percent relative to gasoline (the EPA originally determined corn-ethanol reduces GHG emissions by 20 percent but then revised that including indirect land-use change, and lo and behold, corn-ethanol reduces GHG emissions even more—by 21 percent instead!).[6] Another reason for the higher biodiesel price is that the United States and Brazil have a specific biodiesel mandate where soybean oil is the primary feedstock.

Table 9.1 Alternative measures of biodiesel and ethanol prices

Price ratios[a]	*Values*
Biodiesel over ethanol	
1 United States	1.95
2 European Union	1.82
European Union over the United States	
3 Ethanol	1.32
4 Biodiesel	1.42
United States	
5 Biodiesel MEGD over Ethanol MEGG	1.50
6 Biodiesel MEGD over Ethanol MEGG times 1.34[b]	1.12
7 Biodiesel over free-market biodiesel	1.29
8 Ethanol over free-market ethanol	1.43
9 Biodiesel over free-market biodiesel (adj. for higher diesel tax)	0.77

Notes:
[a] From April 2007 to March 2014.
[b] The 1.34 factor reflects 34 percent more miles per gallon of diesel relative to gasoline.
MEGD, miles equivalent gallons of diesel; MEGG, miles equivalent gallons of gasoline.

Source: US prices from CARD (2014a,b), EU prices from Bloomberg L.P. (2014), and UFOP (2014).

The European Union has an overall mandate but uses canola oil as the principal feedstock, and has a comparative advantage in domestic biodiesel over ethanol production (unlike in the United States and Brazil).

Rows 3 and 4 of Table 9.1 show that the EU price of ethanol (biodiesel) averaged 1.32 (1.42) times that in the United States. We will show later that import tariffs have a lot to do with that, along with the European Union being a natural importer of biofuels.

Despite the sharp increase in the use of biodiesel, there has been no sign of a change in the value of soybean oil relative to the value of soybeans, for example. It is true that the value of soybean oil increased progressively in 2007–2008 and reached a peak of 47 percent in May 2008 but declined rapidly in the Great Recession and has been at normal levels since then. Likewise, there has been no sign of a change in the relative price of rapeseed oil to soybean oil (it peaked at 1.68 in December 2008 but declined to 1.01 in May 2013). Because one obtains more rapeseed oil per acre of rapeseed than soybeans, there has been a switch to more acres of rapeseed, however. For example, according to the data from the Food and Agriculture Organization of the United Nations Statistics Devision (FAOSTAT), the US (global) harvested area of rapeseed between the years 2005 and 2012 increased by 55 (23) percent whereas the area of soybeans increased only by 6.8 (13) percent.[7] Like in the case of corn-ethanol, all crop prices are affected by biodiesel policies through the law of one relative international price of grains and oilseeds.

This chapter aims to provide insights into the functioning of the oilseed–biodiesel–diesel market complex in a large country that determines the biodiesel price and to determine how the market equilibrium changes in response to diesel price shocks. By incorporating the value of joint products, processing costs, fuel taxes, and biodiesel policies, we capture the most important features of this market. We assume a large country in world biodiesel and oilseed (soybean, canola) markets that implements either a biodiesel consumption subsidy (e.g., the US tax credit or excise tax exemption at the pump as in the European Union) or imposes a blending mandate on the share of biodiesel in the final diesel fuel blend. The results discussed in this chapter can also be extended to a variety of other oilseeds that are crushed into oil and meal. The analyses of biodiesel from jatropha or palm oil would require a small adjustment.

The implications of biodiesel policies are the same as that for corn-ethanol discussed so far—higher soybean (rapeseed) oil prices and, thus, oilseed prices. However, unlike the *co-product* dried distillers grains with solubles (DDGS) in corn-ethanol production, which further increases the corn price, the *joint products* of biodiesel production are differentiated and do not return to the same market. Thus, the prices of the joint product soybean (or rapeseed) meal decline as more oilseeds are crushed to make biodiesel, thereby moderating the oilseed price increase due to biodiesel.[8] Therefore, the link between oilseed and biodiesel prices differs from that of corn-ethanol described in Chapter 2. It is the jointness in soybean crushing (in other words, soybean oil and soybean meal) together with the differentiated products that breaks the direct link between the biofuel and its feedstock prices. Producing biodiesel necessarily requires the creation of the joint product soybean meal. Thus, when more soybeans are demanded for biodiesel, more soybean meal is dumped into a market where there has been no shift in demand.

The impact of the tax credit versus a mandate is exactly the same as we modeled it in Chapter 2: higher diesel prices (due to higher crude oil prices) result in higher biodiesel prices under the binding tax exemption (tax credit). Alternatively, biodiesel prices decline with higher diesel prices under a binding mandate. The impact of a diesel price surge on soybean prices when there is more than one oilseed used in biodiesel production is ambiguous for either the tax credit or mandate binding scenarios. This is because canola yields more oil per acre than soybean and it is undetermined a priori how markets for each of the crops that can be converted into biodiesel will respond. We also find that for the same biofuel (ethanol or biodiesel) price increase, corn prices increase more (both in absolute and percentage terms) than soybean prices.

The remainder of this chapter is structured as follows. In the next section, we explain the first key trade barriers that affected the price of biodiesel in the United States, European Union, and, therefore, in the world. The following section explains the key links among biodiesel, soybean oil, and soybean meal prices, analogous to that developed for corn-ethanol in chapter 2. In the section after that, we explain the links between the quantities of biodiesel and soybeans. Then we extend the basic model to include a second feedstock (canola) that yields more oil per hectare and is thus more likely to be produced in the era of biofuels. We show how the inclusion of the second feedstock alters the model's responses to a higher diesel price. In the penultimate section, we compare the efficiency of biodiesel and corn-ethanol and make a case that biodiesel may not be that inefficient, depending on how one measures it. The final section concludes.

The Era of US Biodiesel "Splash and Dash" Exports[9]

Biodiesel is the main biofuel for transport used in the European Union, accounting for approximately 70 percent of the volume of all biofuels consumed in the European Union in 2012 (USDA 2013).[10] This suggests that the developments in the world biodiesel markets are very important not only for the EU biodiesel demand (in other words, being able to secure enough biodiesel to meet the mandate) but also for the EU biodiesel industry as a change in the world price of biodiesel has implications for the profitability of the sector.

An important milestone for the EU (and world) biodiesel markets was the "splash and dash" controversy over US exports of biodiesel. Biodiesel received the blender's tax credit of $1 per gallon for any gallon of biodiesel blended with crude oil-based diesel, regardless of its source (domestic or imported) or amount. A typical blend was a "splash" of 0.1 percent diesel and 99.9 percent biodiesel! The resulting blend was then exported or "dashed" to the European Union. EU biodiesel was also eligible so sure enough, some rapeseed oil produced in the European Union was shipped to the United States, "splashed" with diesel, collected the biodiesel tax credit on over 99 percent of the boatful of now officially diesel "fuel" and then shipped back to the European Union![11]

The EU biodiesel producers argued that the US "splash and dash" program disrupted their profit margins and put most of them out of business as US biodiesel exports represented 90 percent of EU imports. However, de Gorter et al. (2011) argue that the US "splash and dash" program had minimal impacts on EU biodiesel producer profits and that, instead, lower world crude oil prices and EU tax exemptions, as well as increased rapeseed oil prices, negatively affected the profitability of EU biodiesel production. Moreover, the authors maintain that EU biodiesel imports from the United States could be beneficial for EU taxpayers, fuel consumers, and US biodiesel producers.

Upon request from EU biodiesel producers, the European Commission started an investigation into biodiesel imports from the United States in June 2008. (Figure 9.1 shows the key milestones for the German biodiesel industry, a proxy for the EU biodiesel sector.) After completing the investigation, the European Commission found material injury and on March 11, 2009, implemented provisional anti-dumping and countervailing duties on biodiesel imports from the United States. The provisional duties were transformed into five-year duties on July 7, 2009, which were to expire in July 2014.

The tariffs were imposed although the United States closed the "splash and dash" loophole by disallowing US biodiesel imports to be re-exported, thus making only biodiesel produced in the United States eligible for "splash and dash." In the meantime, the US biodiesel blender's tax credit expired at the end of 2009, but has been reenacted retroactively several times after that; the latest reenactment took place in April 2014 and the tax credit was recently extended through 2015. de Gorter et al. (2011) show that the EU import tariffs on US biodiesel did not help the profitability of

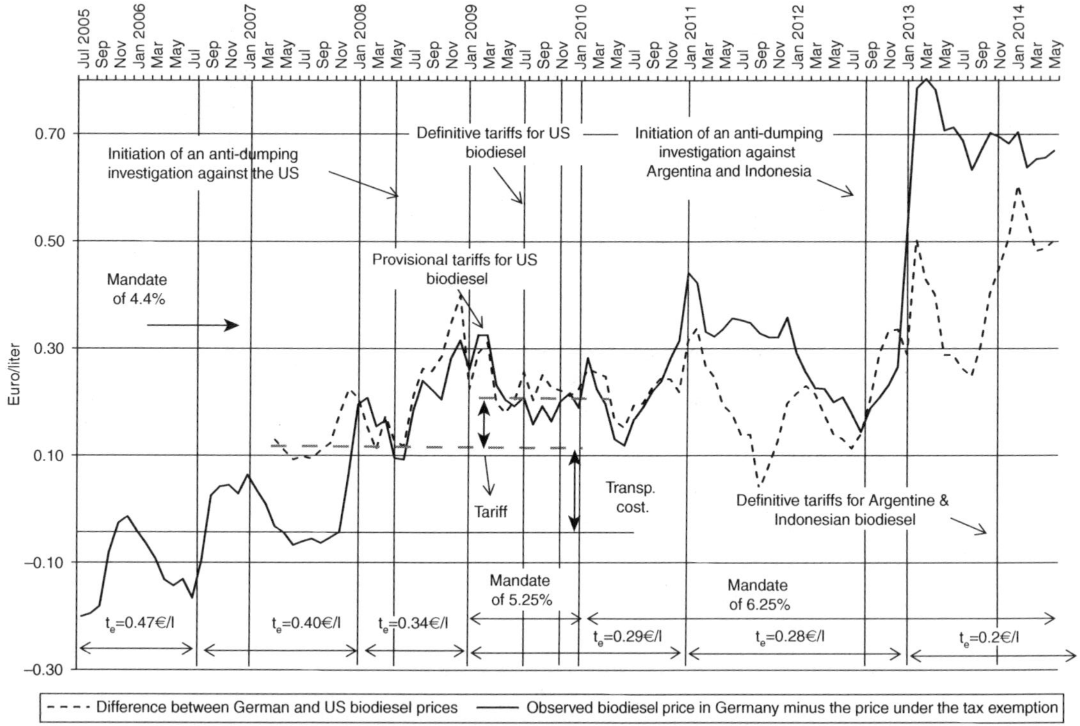

Figure 9.1 German and US biodiesel prices (Euro/liter).

Source: CARD (2014b), UFOP (2014).

the EU biodiesel sector, partly because of increased biodiesel imports from other countries, representing loopholes that allowed US biodiesel to get transshipped, and because mandate premiums were minimal leading into this time period relative to future levels.

In Figure 9.1, we present the difference between the observed German biodiesel price and the predicted price under a tax exemption (using an analogous equation to equation (2.2) in Chapter 2). The figure also depicts the difference between German and US biodiesel prices. One possible reason for the big difference is the imposition of countervailing and anti-dumping duties on the imports of biodiesel from the United States. The imposition of anti-dumping duties on the Argentine and Indonesian biodiesel in 2013 made the difference even bigger. Reduced imports of biodiesel to the European Union contribute to the recent increase in the EU price as well.

Relevant events in the biodiesel market between 2005 and 2014 are labeled in Figure 9.1. Before the imposition of tariffs on US biodiesel exports, the difference between the biodiesel market price in Germany and the United States was equal to transportation and transaction costs: they averaged 0.15 Euro per liter from November 2006 to June 2008. After the tariff was imposed, this difference increased to 0.25 Euro per liter. Subtracting the former figure from the latter, we should obtain approximately the applied tariff rate. However, the difference of 0.10 Euro per liter coincides only with the simple average of the imposed antidumping tariffs of 0.11 Euro per liter (de Gorter et al. 2011), but seems not to reflect the average countervailing duty of 0.19 Euro per liter (Official Journal of the European Union 2009). This can, perhaps, be explained by other countries, such as Argentina, increasing exports directly to the European Union, which implicitly subsidizes biodiesel exports with differential export taxes (Tomei and Upham 2009).

Figure 9.1 also documents that the support to biodiesel in Germany via an excise tax exemption is being phased out and currently is almost zero. However, the difference between the observed biodiesel price and predicted price under a tax exemption has been increasing, suggesting increasing price premiums due to the country's biofuel mandate (6.25 percent in energy equivalent at the present time).

To complement the events depicted in Figure 9.1 with trade data, and to show the severity of both imported and domestic biodiesel using "splash and dash," Figure 9.2 depicts US gross and net imports of biodiesel. The gross imports increased with "splash a dash" (peaked in 2008, which coincides with the peak of US biodiesel exports to the European Union) as, for example, palm oil biodiesel on a boat docked in Texas had to be an official import to get splashed, receive \$1 per gallon tax credit and be shipped to

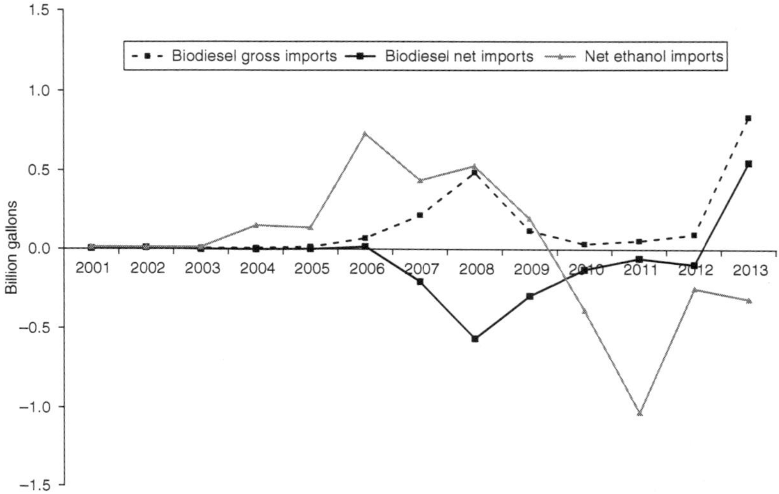

Figure 9.2 US biodiesel and ethanol imports (billion gallons).

Note: See also Chapter 10.

Source: EIA (2014b).

Europe. After the initiation of an anti-dumping investigation, EU imports of US biodiesel declined (see also Figure 9.1). Figure 9.2 also documents that biodiesel exports from domestic supplies are a mirror image of the imports that were exported, perfectly illustrating the gold rush that began in 2006 and ended by 2010 in "splashing and dashing" imported and domestic biodiesel to get benefits from both the US tax credit and high EU biodiesel prices due to EU policies.[12] We will refer back to Figure 9.2 in the next chapter because it illustrates how the United States has become a net exporter of ethanol and a net importer of biodiesel in recent years, all having to do with the ethanol blend wall and the interaction of various sub-mandates in the complex structure of the US biofuel mandate.

Another example of the complexity of a tax credit versus tax exemption is Canada that switched its tax exemption to a biofuels producer tax credit in April 2008 (Auld 2008). Because Canada is too small to influence either the world market price of ethanol, corn, or oil, the former tax exemption in Canada subsidized domestic fuel consumption (ethanol and gasoline). However, the revised Canadian biofuels producer credit program subsidizes biofuels production instead (and Canada had exported all domestic biodiesel production, taking advantage not only of the domestic production subsidy but also of both the high market price established by the EU tax exemption and the US blender's tax credit). Until the United States closed the "splash and dash" loophole, Canada had, therefore, "triple dipped": it benefitted from its own biofuels producer tax credit (available only for domestic production), the US blender's tax credit (available to US production and US imports), and a high EU biodiesel price due to EU biodiesel excise tax exemptions.

Although the European Union implemented countervailing and anti-dumping tariffs on the imports of biodiesel from the United States, biodiesel coming from Argentina and Indonesia was not subject to these duties implemented in July 2009. Four years later, after essentially blocking biodiesel imports from the United States, the European Union imposed anti-dumping duties also on biodiesel from Argentina and Indonesia in order to protect the domestic market from imports of subsidized biodiesel from these countries (e.g., Argentina has used a system of differential export tariffs favoring exports of biodiesel to exports of biodiesel feedstocks). As a result of barring the EU market for biodiesel from the major producers, Argentina, the world's largest biodiesel exporter, diverted some of its biodiesel deliveries to the United States, which has become a net importer of biodiesel since 2012 and is currently relying on biodiesel to overcome the ethanol blend wall (explained in the next chapter).

In the meantime, the growing demand for biodiesel in the European Union combined with insufficient domestic production and loss of foreign sources open a window of opportunity for the US biodiesel for which the anti-dumping import duties were to expire (after five years) in July 2014. However, whether the US biodiesel will be shipped to the European Union is uncertain at this point for at least two reasons. First, as we will show in the next chapter, biodiesel helps to overcome the US blend mandate. Second, should the US biodiesel exporters claim the $1 per gallon tax exemption, which they inevitably will, the European Union would likely prolong the anti-dumping duties.

The Link Between Biodiesel, Soybean Oil, and Soybean Meal Prices

We begin the analysis on biodiesel derived from soybeans. The economics of rapeseed biodiesel is identical though the model needs to be adjusted somewhat when palm oil is examined. We extend the model to include canola oil in the latter part of this chapter.

The distinguishing feature of the soybean oil-biodiesel market is the *jointness in production*: 1 bushel of soybean (60 lbs) when crushed yields β_1 = 11.28 lbs (19 percent of weight) of oil and β_2 = 48.72 lbs (81 percent) of meal (FAPRI 2014). Its high-protein content makes the soybean meal an important component in feed for hogs, chickens, and cattle. We assume that processors of soybeans operate under constant returns-to-scale and that the processing cost c_{0s} (crushing margin) per pound of soybean oil is constant. Denoting the price of soybean oil and meal as P_{SO} and P_M, respectively, the revenue from one bushel of soybean crushed is equal to $\beta_1 P_{SO} + \beta_2 P_M$. On the other hand, the cost of jointly producing the soybean oil and meal is $P_{SB} + \beta_1 c_{s0}$, where P_{SB} denotes the market price of purchased soybeans. A competitive processor will crush soybean until the marginal revenue and the marginal cost per bushel of soybean are equal. Thus, the market price of soybean can be expressed as

$$P_{SB} = \beta_1 P_{SO} + \beta_2 P_M - \beta_1 c_{0_s}. \tag{9.1}$$

Soybean oil can be used either as an input for biodiesel production or for human consumption.[13] One pound of soybean oil yields $\beta_3 = 0.13$ gallons of biodiesel.[14] Akin to soybean crushing, competitive biodiesel producers will equilibrate the revenue earned from selling biodiesel produced per pound of soybean oil, $\beta_3 P_B$, with the total cost of purchasing one pound of soybean oil and processing it into biodiesel, $P_{SO} + \beta_3 c_{0b}$, where P_B denotes the biodiesel market price and c_{0b} denotes the processing cost per gallon of biodiesel. The link between prices of soybean oil and biodiesel can thus be written as[15]

$$P_{SO} = \beta_3 (P_B - c_{0b}). \tag{9.2}$$

Equation (9.2) is the key link between biodiesel and soybean prices, and Figure 9.3 shows that it is an excellent predictor of the US soybean oil price except for the years 2011 and 2013.[16] In those years, the model severely over predicts soybean oil prices. The reason is excess profits enjoyed by biodiesel producers in those years, which may seem surprising as there is officially significant excess capacity in biodiesel production in both the United States and European Union.[17] However, unexpected market developments in each of 2011 and 2013 caused these excess profits as biodiesel plants were unable to increase production (and, thus, bid up soybean oil prices) fast enough to eliminate these excess profits. The reasons for these over predictions will be explained in more detail in the following chapter as it involves the interaction of the nested US mandate structure. Briefly, in 2011 ethanol prices in Brazil skyrocketed and biodiesel was required to fill the mandate for advanced biofuels on short notice (for which corn ethanol is not eligible) whereas in 2013, because of the ethanol blend wall (see next chapter), biodiesel was called upon on short notice to not only fill the entire mandate for advanced biofuels but also to contribute to the residual component of the US mandate where corn-ethanol is eligible.

We re-do the entire exercise above for rapeseed oil in the European Union, using prices and taxes for diesel and biodiesel in Germany as a proxy for the European Union.[18] In contrast to US soybean oil prices, the prediction error for the German rapeseed oil price (the dashed line in the lower part of Figure 9.3) is significantly higher, especially for the past two years.

The model over predicts rapeseed prices more often than not, suggesting that the German rapeseed oil processing industry made profits in producing biodiesel. The only exception is a short period between April 2012 and October 2012 when the predicted rapeseed oil price is lower than the observed one. Since July 2012, the gap between the predicted and actual rapeseed prices in Germany has been widening. The beginning of this period coincides with the initiation of anti-dumping investigation for Argentina and Malaysia.[19] The reason for the over-prediction is excessively high biodiesel prices relative to what they normally are. Biodiesel prices averaged $6.42 per gallon in 2013, far higher than US biodiesel prices or German biodiesel prices in 2011 of $4.55 per gallon.

The distinguishing feature of the price links in equations (9.1) and (9.2) is that they are tiered: an exogenous shock in the biodiesel price first affects the price of soybean oil and then the price of soybeans, accounting for the decline in the price of soybean meal. This makes the link between biodiesel and soybean (or rapeseed) prices less direct than the link between ethanol and corn prices. Therefore, the increase in the soybean price due to an

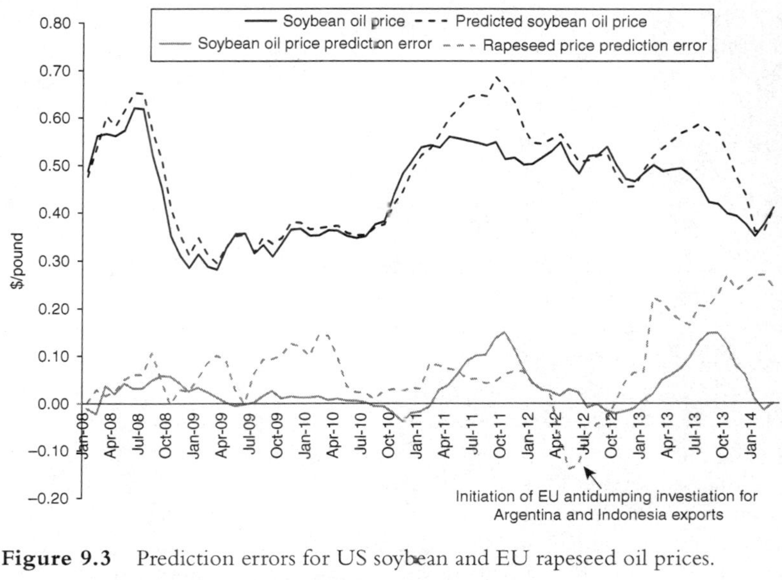

Figure 9.3 Prediction errors for US soybean and EU rapeseed oil prices.

Source: Calculated using data from CARD (2014b) and UFOP (2014).

increase in the biodiesel price is expected to be smaller than the percentage increase in the corn price for the same shock in the ethanol price.

To illustrate the difference between corn-ethanol and soybean oil-biodiesel, let us consider the marketing year 2012–2013 in which the average price of soybean and corn was \$14.40 per bushel and \$6.89 per bushel, respectively.[20] Consider an exogenous increase in the price of both biofuels of \$0.10 per gallon. In doing this back-of-the-envelope calculation, we assume the price of soybean meal does not respond to the change in the biodiesel price (thus, we overestimate the change in the soybean price). The expected increase in the soybean price is $11.28 \times 0.13 \times 0.10 = \$0.15$ per bushel while the predicted increase in the corn price is $4 \times 0.10 = \$0.40$ per bushel. The coefficient of four is the average corn-ethanol price multiplier for 2012–2013. Thus, the same shock in the price of a biofuel has a greater effect (in absolute terms) on the price of corn than on the price of soybean. Given the 2012–2013 price values, these absolute price changes translate into a 1- and 5.8-percent increase in the soybean and corn prices, respectively.

However, the initial levels of biofuels and commodity prices will affect these calculations. To eliminate this effect, we calculate price transmission elasticities for ethanol-corn and biodiesel-soybean prices by dividing the percentage change in the price of a commodity by a percentage change in the price of a biofuel. In the marketing year 2012–2013, the average biodiesel and ethanol prices were \$4.55 per gallon and \$2.38 per gallon, respectively. Therefore, a \$0.10/gallon price increase translates into 2.20 and 4.20 percent for biodiesel and ethanol prices, respectively. The estimated price transmission elasticity for biodiesel is then 0.46 (=1/2.2), whereas for corn-ethanol it is about three times higher at 1.38. The corn price response to an ethanol price shock is greater than the response of the soybean price to a biodiesel price shock.

The intuition for this result stems from the jointness in soybean oil production. A higher biodiesel price drives up the price of soybeans which results in more soybean production, and, thus, more soybeans crushed which leads to more soybean meal produced. In order for the market to clear, the price of soybean meal has to decrease (or at very least remain the same), offsetting the soybean price rise [as per equation (9.1)]. On the other hand, a higher ethanol price attracts more corn to ethanol production, reducing the amount of corn for food and feed use. This makes the DDGS relative price increase, thus reinforcing the increase in the corn price [as per equations (2.1) and (2.1′) in Chapter 2].

Table 9.2 presents estimates of the minimum corn and soybean prices when the commodities are directly linked onto crude oil prices (whose values are specified in the first column). Based on simple regressions of gasoline on crude oil prices (and also for diesel) using historical prices, the

Table 9.2 Predicted gasoline/diesel prices for alternative crude oil prices and the lowest corn and soybean prices can go

	Predicted price ($/gallon)		*Lowest prices can go*			
Brent crude oil price ($/barrel)	*Gasoline*[a]	*Diesel*[b]	*Ethanol ($/gallon)*	*Biodiesel ($/gallon)*	*Corn ($/bushel)*	*Soybean ($/bushel)*
80	2.14	2.26	1.35	2.01	2.44	5.55
85	2.25	2.39	1.43	2.13	2.77	5.98
90	2.37	2.52	1.51	2.25	3.09	6.41
95	2.48	2.64	1.59	2.37	3.42	6.85
100	2.60	2.77	1.67	2.48	3.74	7.28
105	2.71	2.90	1.75	2.60	4.07	7.71
110	2.83	3.03	1.83	2.72	4.39	8.15
115	2.95	3.16	1.91	2.83	4.72	8.58
120	3.06	3.28	1.99	2.95	5.04	9.02
125	3.18	3.41	2.08	3.07	5.37	9.45
130	3.29	3.54	2.16	3.18	5.70	9.88
135	3.41	3.67	2.24	3.30	6.02	10.32
140	3.52	3.80	2.32	3.42	6.35	10.75
145	3.64	3.92	2.40	3.53	6.67	11.18
150	3.76	4.05	2.48	3.65	7.00	11.62

Note:
[a] Gasoline price = 0.286 + 0.023 × Crude oil price.
[b] Diesel price = 0.212 + 0.026 × Crude oil price.
The regression coefficients were estimated by OLS using prices of Brent crude oil, and gasoline and diesel US Gulf Coast prices from EIA (2014c).

Source: CARD (2014a, b), EIA (2014a), GaspriceWatch.com Federal and State Fuel Taxes.

second and third column predicts gasoline and diesel prices corresponding to various levels of the crude oil price. In the fourth column, using equation (2.2) of Chapter 2, we estimate what the ethanol market price would be if no biofuel policy were in place (a similar equation is used to determine the biodiesel price under free markets in the fifth column—this equation (9.4) is presented and discussed below). The prices of both biofuels increase with a higher crude oil price because consumers are willing to pay more on a miles traveled equivalent basis when the crude oil price rises. Finally, the last two columns give the floor for corn and soybean prices under the assumption that no biofuel policy is in place but that the biofuel-feedstock zero profit condition holds.

There are a couple of things one should take note of in Table 9.2. First, the relative price of soybeans over corn in Table 9.2 (in $ per bushel) averages 1.87 (historically, soybean prices are approximately 2.4 times that of corn). Clearly, the "lowest price" that corn and soybeans prices can each go is inconsistent with the law of one relative international price of grains and oilseeds. Therefore, in the "real world," one price has to be higher than the lowest it can go to maintain equilibrium (or even go below it—see the epilogue for such a possible scenario). Second, the relative price of soybeans decline as crude oil prices increase. Because our predictions in Table 9.2 are based simply on one equation without having market adjustments, for example, changes in the prices for DDGS and soybean meal, the model can be thought of only as a first-round effect. However, it does show that for a particular per dollar a gallon change in the biofuel price, biodiesel has a smaller impact on crop prices than ethanol.[21]

The Link Between Biodiesel and Soybean Quantities

Because biodiesel is not directly produced from soybeans, unlike the case for ethanol from corn, the derivation of the biodiesel supply curve is more complex. Note that equation (9.2) implies a one-to-one relationship between biodiesel and soybean oil prices. Because the quantity of biodiesel can be converted into soybean oil (through the coefficient β_3), it means that the derivation of the biodiesel supply curve is equivalent to the derivation of the excess supply curve of soybean oil for biodiesel.

To focus on the key relationships between soybean and biodiesel markets, let us assume that the demand for soybean meal is perfectly elastic, meaning that the market can absorb any amount of meal without affecting its price. Under this assumption, equation (8.1) becomes

$$P_{SB} = \beta_1 P_{SO} + k, \tag{9.3}$$

where k is a parameter capturing all exogenous parameters in equation (9.1). Equation (9.3) implies a one-to-one relationship between prices of soybean and soybean oil.

Consider first a situation where no biodiesel is produced. Using equation (9.3) and the quantity conversion factor β_1 (pounds of soybean oil from one bushel of soybeans), we can convert the demand for soybean oil for human consumption into an equivalent demand curve for soybeans. The market price of soybean is then determined by the intersection of the soybean supply curve and this converted demand curve. We denote this resultant price as P_{SBN}—the non-biodiesel soybean price (a counterpart to P_{NE} discussed in Chapter 2). At any soybean price above P_{SBN}, there will be excess supply of soybeans—the horizontal difference between soybean supply and demand curves—that can be processed into soybean oil for biodiesel production. Using equation (8.3) and the parameter β_1, we can convert the excess supply of soybeans into an excess supply of soybean oil for biodiesel production.

Finally, to arrive at the biodiesel supply curve, we use equation (9.2) to determine prices and the parameter β_3 (gallons of diesel from one pound of soybean or rapeseed oil) for quantities and apply them to the excess supply curve of soybean oil. Corresponding to the soybean price P_{SBN} is the biodiesel price which determines the intercept of the biodiesel supply curve. Because the intercept is positive, it is likely that there will be "water" in the biodiesel policy price premium (as was the case with corn ethanol in Chapter 2). The likelihood of the occurrence of "water" depends on (1) the position of the intercept of the biodiesel supply curve and (2) the free-market biodiesel price. The magnitudes of the both factors need to be determined empirically.

As in equation (2.2) in Chapter 2, the free-market biodiesel price, P_B^*, is given by

$$P_B^* = \gamma P_D - (1-\gamma)t_D, \tag{9.4}$$

where $\gamma = 0.913$ denotes miles traveled per gallon of biodiesel relative to a gallon of diesel, P_D denotes the market price of diesel and t_D denotes the diesel tax. The sum of the US federal and (average) state diesel tax in 2013 was \$0.55 per gallon (\$0.495 per gallon for gasoline). In the same year, the US diesel price averaged \$2.97 per gallon.[22] This translates into a free-market biodiesel price of \$2.45 per gallon. Because the diesel price has historically been above the gasoline price; the relative mileage for biodiesel is greater than for ethanol; and because the penalty to biodiesel blenders [in other words, the term $(1 - \gamma)\ t_D$] is smaller than for ethanol, the free-market biodiesel price will be higher than the corresponding price for ethanol. Other things held constant, this curbs the "water" (if any) in biodiesel policies as compared to ethanol.

The Economics of Having Two Feedstocks for Biodiesel (Soybean and Canola Oil)

Because of growing biodiesel demand, the increasing value of oilseed oil incentivizes the switch to more oil-intensive crops such as canola (as we showed in the introduction). The increasing demand for biodiesel is reflected in the biodiesel market prices. US biodiesel prices went from a low of \$2.68 per gallon in March 2009 to a high of \$5.45 per gallon in July of 2011 (the low in January 2014 was \$3.34 per gallon). Later, the US biodiesel prices have come down from the period of excess profits (averaged \$4.73 per gallon from January to October 2013 and \$3.70 per gallon from November 2013 to March 2014).[23]

Because oil represents approximately 40 percent of the weight of canola[24] and only approximately 20 percent of soybean,[25] more oil can be obtained from one hectare of canola. This has implications for developing countries as they decide between (or may switch to) soybeans, palm oil, or jatropha.[26] Canadian and US farmers are converting to canola as the oil content of canola is much higher, although the yield per acre is lower than that for soybeans. Because of food demand for oils, one may not expect a high divergence in soybean oil and canola oil prices but crop prices could diverge where canola prices increase relative to soybeans.

Drabik et al. (2014) model the market effects of an exogenous increase in the diesel price in an environment where biodiesel is produced from soybean and canola oils. Furthermore, they analyze these market effects separately for a binding tax exemption (tax credit) and a binding mandate. When the tax exemption is binding, a higher diesel price makes the canola price rise but the effect on the soybean price is generally ambiguous, depending on the elasticities of meal demand and canola supply.

To see when this perhaps counterintuitive result can occur, consider the extreme case when the demand for oilseed meal is perfectly inelastic. In the periods when the tax exemption is binding and diesel prices are high, farmers will tend to switch to canola as it contains more oil and hence more biodiesel can be produced. But more canola crushed, everything else held constant, means more meal produced. As the demand for meal is assumed to be inelastic, the increase in the amount of meal from canola has to be offset by a reduction in the amount of meal coming from soybeans which necessitates a reduction in soybean production, and a corresponding drop in its price. However, Drabik et al. (2014) find empirically that prices of both canola and soybean increase with a higher diesel price. This occurs because the demand for meal is not perfectly inelastic; thus, there is room to accommodate more meal coming from higher oilseed production.

In the case when a mandate is binding, an increase in the diesel price has opposite effects on the market equilibrium as compared to a binding tax exemption (tax credit). Moreover, for the same biodiesel production, a shock in the diesel price under a binding blend mandate results in a change in canola (soybean) price of a lower magnitude than under a binding tax exemption. The canola price unambiguously decreases under the mandate while the soybean price tends to decrease with more elastic canola supply and a less elastic demand for oilseed meal.

The empirical results of Drabik et al. (2014) illustrate how the biodiesel and feedstock markets respond to an exogenous diesel price shock under alternative biofuel policies. These results can be summarized as follows. First, the market prices respond much more under a binding tax exemption (tax credit) than under a mandate, regardless if biodiesel is produced from soybean oil only or also from canola oil. For example, a 10-percent increase in the biodiesel price in Drabik et al. (2014) causes the soybean price to increase by approximately 4.5 percent under the tax exemption, but to decrease by a negligible 0.01 percent under the mandate. The intuition for the latter result is that the demand for diesel fuel is empirically very inelastic; therefore, the consumption of diesel fuel changes minimally due to higher diesel prices, leaving the biodiesel, and thus oilseed prices, essentially unchanged. With a binding tax exemption, a higher diesel price has a direct impact on the feedstock market prices, however.

Second, consistent with the qualitative prediction, a diesel price shock has opposing effects on market prices: the prices of feedstocks and oils increase, and the price of meal decreases under the tax credit, whereas the opposite happens with the mandate. Third, the market effects of an exogenous diesel price increase are essentially identical for soybean, soybean oil, and meal in the models with and without canola.

Is Biodiesel Efficient Compared to Ethanol?

Many economists have argued that biodiesel is a very inefficient biofuel compared to corn- or sugarcane-ethanol. The reputation of biodiesel being an inefficient biofuel may be because of the high tax credit of $1 per gallon (almost twice that of the former US ethanol tax credit), or due to the excess production capacity in both Europe and the United States or that diesel is a secondary transportation fuel relative to gasoline in the United States. Furthermore, it is easy to conclude that biodiesel is inefficient because prices per gallon averaged almost twice that of ethanol in both the United States and Europe in our discussion of Table 9.1 earlier.

However, it is difficult to determine which biofuel is more efficient. One way to determine efficiency would be to compare the price of biodiesel to that of ethanol and adjust for the fact that a gallon of biodiesel averages 0.91 miles relative to a gallon of diesel compared to 0.70 miles for ethanol relative to gasoline. If one adjusts for this differential, then the ratio of US biodiesel to ethanol prices on a miles traveled per gallon adjusted basis falls to 1.5. If one then adjusts for the fact that a gallon of diesel averages 34 percent more miles traveled than a gallon of gasoline, row 6 in Table 9.1 shows that the ratio of biodiesel over ethanol prices is 1.12, close to one. Therefore, can one argue biodiesel is close to being efficient vis-a-vis ethanol? The next two rows in Table 9.1 provide the ratio of biodiesel and ethanol prices to their respective free-market levels. Here, the distortion in the ethanol prices is higher. What can we conclude from this information?

The final row in Table 9.1 gives the biodiesel price over the free-market price of biodiesel, adjusted for the fact the motor diesel fuel tax should be

66 percent of that of motor gasoline because diesel gets 34 percent more miles (therefore, we set the motor diesel tax equal to 66 percent of the actual motor gasoline tax). Under this scenario, the observed biodiesel price is below that of the free-market biodiesel price. However, much more research has to be done to determine the economics of biofuels and policies affecting biofuels like fuel taxes and biofuel policies.

Concluding Remarks

With a 20-percent and growing share of the global market, biodiesel is the world's second most produced biofuel. Biodiesel plays a key role in fulfilling the EU biofuel mandate but is also enjoying growing importance in the United States where, as an advanced biofuel, we will show in the following chapter that it is helping overcome the ethanol blend wall.

Because the European Union has historically been a net importer of biodiesel, the EU biodiesel industry together with the EU Commission have closely been watching the developments in biodiesel policies implemented worldwide. For example, the European Union has already imposed anti-dumping import tariffs on biodiesel coming from the United States (in March 2009) and Argentina and Malaysia (in November 2013). The imposition of tariffs for the US biodiesel was not successful in improving the economic standing of the EU biodiesel sector because of transshipments of US biodiesel through other countries and imports from countries not subject to the anti-dumping duties (e.g., Argentina).

The main crop for biodiesel production in the United States is soybeans whereas it is rapeseed in the European Union. On the other hand, Malaysia and Indonesia derive biodiesel from palm oil. We show how the jointness in soybean (and rapeseed) crushing—in other words, the production of both soybean oil and meal—changes the otherwise direct link between a biofuel and its feedstock (e.g., the case for ethanol and corn). More specifically, an exogenous shock in the biodiesel price has a direct effect on the soybean oil price but does affect the soybean price through a change in the soybean oil price accompanied with a (opposite) change in the price of soybean meal. This is the reason why the response of the soybean price to a biodiesel price shock is smaller (both in absolute and relative terms) than the corn price response to an equal ethanol price shock. In addition, the price transmission elasticity of an ethanol price shock to the corn price is higher than a corresponding elasticity for biodiesel and soybeans.

We also find that whereas higher biodiesel prices make canola (a type of rapeseed) a more preferred choice over soybeans (because of more oil extracted per hectare of oilseed planted), the net effect of a higher diesel price on soybean is ambiguous, depending on the relative elasticity of the meal demand and canola supply.

Although it has been argued that biodiesel is a very inefficient biofuel compared to corn- or sugarcane-ethanol, we show it does not always hold and the result depends on whether or not one adjusts for the relative mileage of both biofuels and the relative mileage of diesel versus gasoline.

The theoretical results discussed for a large country produce several interesting implications for small developing countries, such as Zambia or Senegal, that are price takers in all relevant commodity markets and that may be considering the promotion of biodiesel production (Msangi and Evans 2013). First, because a higher diesel price increases the biodiesel market price under a binding tax exemption but reduces it under a binding mandate, small developing countries should closely analyze which policy determines the biodiesel price in large countries (e.g., represented by OECD countries). This advice becomes even more relevant in the periods of unstable crude oil prices (Drabik et al. 2014).

Second, because in the new biofuel era crude oil prices not only affect biodiesel prices but also feedstock prices, farmers in small developing countries should carefully plan the pattern of production of biodiesel feedstocks (e.g., soybean versus canola), keeping in mind that their world prices, and, thus, profitability, will not only depend on the binding biodiesel policy in the biodiesel price determining country, but also on the relative elasticities of meal demand and canola supply. This uncertainty about the feedstock prices might adversely affect the stability of feedstock supply in developing countries, thus affecting the profitability of the biofuel sector in those countries (Msangi and Evans 2013).

CHAPTER 10

THE COMPLEX STRUCTURE OF THE US BIOFUEL MANDATE: A HANDBOOK

with Justin Shepherd

Taleb (2012, 373) notes that "We have developed a fondness for neomanic complication over archaic simplicity." Never have truer words been spoken when describing and explaining the logic of the US biofuel mandate under the rubric of the renewable fuel standard (RFS) with its nested structure and many interactions. For example, we delayed our explanation for the prediction errors in 2011 and 2013 of the soybean oil–biodiesel price linkage model because it hinges on the interaction of sub-mandates in the RFS. It turns out that the nested structure of the US mandate also has implications for corn-ethanol and corn prices, and trade in both ethanol (including two-way trade with Brazil) and biodiesel. Therefore, prices of both biofuels and three main feedstocks: corn, soybean, and canola oil and sugarcane (and, thus, sugar) are being affected by the way in which the complex US RFS mandate works. Furthermore, the primary driver of prices and trade since early 2013 has been the ethanol blend wall in the United States and more recently, uncertainty in what the Environmental Protection Agency (EPA) will finally rule for 2014 (which at the time of writing, has still not been resolved),[1] both having implications for the way forward in 2015 and beyond. Therefore, we devote an entire chapter to explain the economics of the complex structure of the US mandate, given its importance, and as the reader will quickly find out, its complexity. However, the volatility of grain-oilseed prices have been greatly impacted by this mandate structure in 2011 and now again since 2013; therefore, patience is asked of the reader to bear with us as we plough through all of this.[2]

You might also ask why the structure of the US mandate is so unique compared to, say, that in the European Union or Brazil. Well, the policy structure impacting the Brazilian ethanol market is also uniquely complex, the details of which will be explained in the very next chapter. However, the EU biofuel mandate is rather straightforward: an overall quantity target (inexactly adjusted for differential mileage between ethanol and biodiesel)[3] with each country being responsible for its implementation, with many countries not meeting the target and with each country presumably having their own time schedule. The EU policies have been dominated by actions to reduce the excise tax exemptions at the retail fuel outlets and the imposition of many and high import barriers for both ethanol and biodiesel, especially anti-dumping and countervailing duties. Sustainability standards have also hindered imports of biofuels in to the European Union and may be illegal under World Trade Organization (WTO) rules (Switzer and McMahon 2011).

This chapter is outlined as follows. We begin by briefly reviewing the EPA's rulemaking regarding the RFS and how different biofuels are classified within the mandates. We will show how the four mandate categories interact and the particular spill over pathways that exist. In addition, the EPA is required to implement the annual quantities for each of the four mandates defined earlier by specifying blend ratios for the amount of renewable fuel blended with each of gasoline and diesel. We look at different methods of calculating ethanol and biodiesel blend ratios and the confusion that stems from the RFS structure. We also look at how renewable identification numbers (RINs) are utilized for compliance of each blender's RFS obligations and differing gallons of ethanol equivalent values that lead to certain biofuels being preferred. Finally, we consider why 2013 is an important year for the ethanol blend wall (the 10-percent ethanol limit in gasoline) resulting from the RFS and how the market accordingly priced RINs such that imported biodiesel is breaching the blend wall instead of increased E85 sales.

The Basic Structure of the US Mandate

US biofuel mandates under the rubric of the RFS are extremely complex.[4] Lewis (2014, 198) remarks that "*People think that complex is an advanced state of complicated," said Zoran. "It's not. A car key is simple. A car is complicated. A car in traffic is complex.*" We show this to be the case amongst the confusion of understanding the RFS. As outlined in the first column of Table 10.1, four broad categories of biofuel mandates are specified: cellulosic biofuels,[5] biomass based diesel (see Box 10.1), advanced biofuels, and the overall or total "mandate" (hereafter defined as the RFS). The cellulosic biofuels

Table 10.1 Classifying biofuels in the United States

Biofuel type	*GHG reduction*[a]	*Equivalence value*	*RIN category*[b]
Cellulosic Biofuel Mandate	60%		
Biogasoline		1.5	D3
Biodiesel		1.7	D7
Biomass-based Diesel Mandate	50%		D4
Biodiesel		1 (1.5)[c]	
Renewable diesel[d]		1 (1.5–1.7)[c]	
Advanced Biofuel Mandate			
Residual or undifferentiated advanced biofuels	50%		D5
Sugarcane ethanol		1	
Naphtha		1.5	
Biobutanol		1.3	
Biogas		1	
Total RFS (residual or renewable fuel gap)	20%		D6
Corn Ethanol		1	

Notes:

[a] Biofuels within mandate must reduce GHG by that amount over either diesel or gasoline.

[b] Renewable Identification Number is the way in which the EPA ensures compliance in each category, discussed later in the chapter.

[c] Numbers in parentheses reflect equivalence values for the advanced and total RFS mandates.

[d] Non-ester renewable diesel uses a different production process than biodiesel, the latter using the esterification process.

Source: U.S. Government Printing Office. (2014).

and biomass based diesel mandates are nested within the advanced biofuel mandate with the latter nested within the total RFS. The residual of the advanced mandate is often referred to as "undifferentiated advanced biofuels" whereas the residual of the total RFS is often referred to as the "renewable fuel gap."

Each category is defined by two criteria: biomass based diesel (see Box 10.1)[6] versus not (mostly ethanol), and a minimum level of greenhouse gas (GHG) emissions reduction relative to the fossil fuel (gasoline or diesel) it is assumed to replace (see the second column in Table 10.1).[7] For example, cellulosic biofuels are required to have lifecycle GHG emissions that are at least 60 percent less than gasoline or diesel, depending on the type of biofuel while corn ethanol only qualifies for the RFS or renewable fuel gap and are required to reduce GHG emissions by 20 percent relative to gasoline.[8] Eligible biofuels are not only biodiesel and ethanol, but also include renewable diesel, naphtha, biobutanol, biogas, and biogasoline. There is no renewable gasoline counterpart to renewable diesel. We assume all of these alternative biofuels are blended with either gasoline or diesel. Note also that all listed biofuels qualify for the total RFS.

Box 10.1 What is the definition of a biofuel and what is the difference between biodiesel and renewable diesel?

So far, we assumed a biofuel was used for transportation fuel only and is exclusively made up of ethanol and biodiesel. However, not all biofuels are used for transportation. Biogas, cellulosic biodiesel (heating oil), and naphtha are not used as transportation but count toward the mandate (their levels to date have been minuscule and, thus, can be ignored). Furthermore, we now introduce the term "biomass-based diesel," which has its own mandate and is made up of two fuels: biodiesel and renewable diesel. What is the difference between the two, given that they are often made from the same feed stocks (Yoon 2011)? The difference is the production process for each biofuel. Biodiesel is made from transesterification whereas renewable diesel is made by at least three other processes. Unlike biodiesel, renewable diesel has the same chemical properties as diesel. Renewable diesel equivalence values (see discussion in text below) vary between 1.5 and 1.7, depending on the inputs used to produce it. Biodiesel only gets an equivalence value of 1.5. However, in later chapters, we revert back to using the term "biodiesel" to include renewable diesel.

The third column of Table 10.1 provides the "equivalence value" for each biofuel. The numeraire is one and is assigned to both corn- and sugarcane-ethanol along with biogas. All other biofuels get a higher equivalence value. It should be noted that biomass-based diesel has an equivalence value equal to one when counted toward its own mandate but then the value rises to 1.5 for biodiesel (and possibly to 1.7 for renewable diesel) when counted toward the advanced mandate or toward the advanced and RFS mandate residuals (the undifferentiated advanced biofuels and renewable fuel gap, respectively) should biomass-based diesel consumption exceed its mandate.

The cellulosic mandate has equivalence values for each biofuel to fill their own mandate and be automatically counted toward the advanced mandate (and to the residuals for the advanced mandate should they go over their own mandate). Therefore, biomass-based diesel biofuels are unique in that if their mandate is exactly filled (with no overage, which was the outcome for several years), then 50 percent of their consumption (taking biodiesel as an example) is automatically counted toward the advanced biofuel mandate residual, displacing, say, sugarcane ethanol that is a competitor for advanced biofuels.

The final column of Table 10.1 gives the RIN category assigned to each biofuel. These are used for compliance and are explained in Box 10.2 for now and explained in more detail as we discuss other topics in this chapter further.

Before moving on to the next section, let us first summarize what we have learned by reviewing panel (a) of Figure 10.1. The outer ring is the total RFS. The next ring is the advanced mandate. Within that mandate lie two sub-mandates: the biomass-based diesel mandate (with its automatic 1.5 "overage factor") and the cellulosic biofuels mandate (which was zero up until 2014). The advanced mandate residual (or "undifferentiated

Box 10.2 What is a RIN?

Renewable identification numbers ensure blenders fulfill the mandated blend requirements for each category of the total RFS (see final column of Table 10.1 for the various RIN categories and "D" number assigned to it; each "D" number can have different equivalence values, depending on the biofuel). Each gallon of ethanol (biomass-based diesel) is assigned a RIN that is separated at the end of the supply chain when blended with gasoline (diesel), and the RIN is forfeited to the US Environmental Protection Agency as proof of compliance with the RFS. If some blenders blend more ethanol (biomass-based diesel) than the mandated minimum, the ensuing excess RINs can be banked for later use (in lieu of not blending the required minimum) or traded for use by other blenders. RINs can also be borrowed from the following year.

Prior to 2013, US corn-ethanol D6 RIN prices were very low (typically 1–2 cents per gallon) reflecting differential RFS compliance costs. For example, a blender in some remote region of the country minimizes costs by buying RINs from a blender elsewhere who is willing to blend more ethanol than the required minimum. However, RIN prices sometimes rose to double-digit values (to a maximum of approximately 15 cents per gallon) when ethanol prices approached or where above gasoline prices, which gave blenders the incentive to use less ethanol and more gasoline as consumers do not have choice at the pump between pure gasoline and blended fuel (see Figure 10.3). No consumer choice means blenders have the option to "water down the scotch," but if ethanol prices exceed gasoline prices, then there is an incentive to blend only the minimum amount of ethanol required.

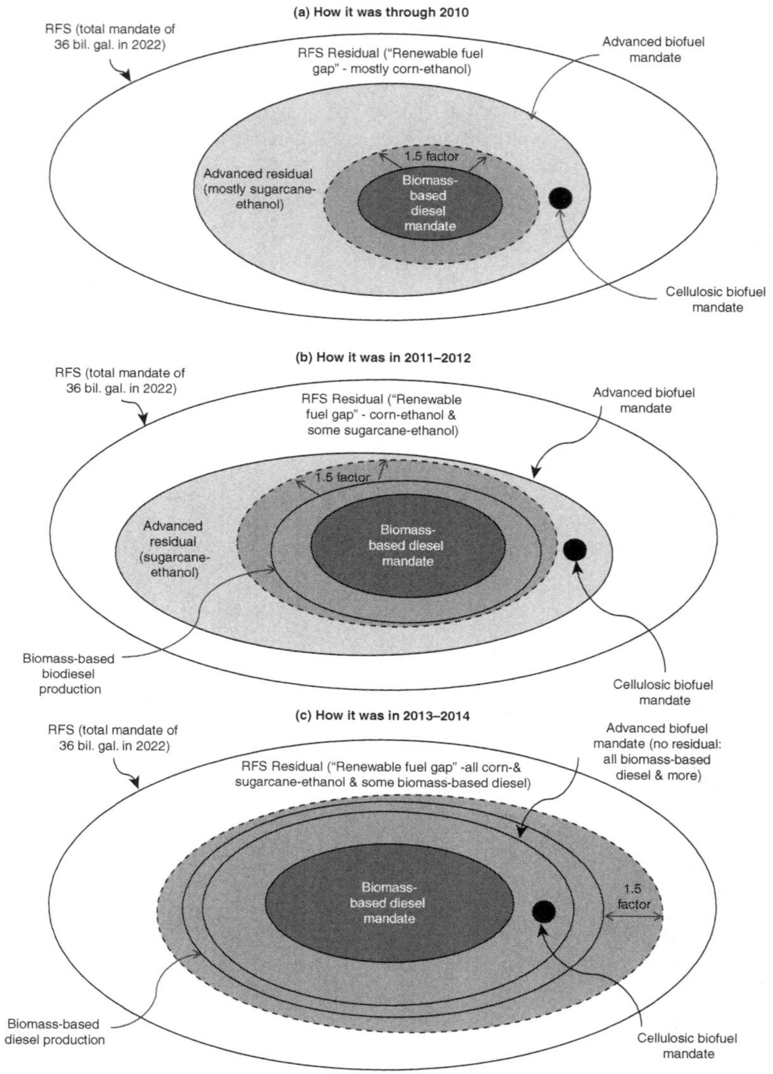

Figure 10.1 The nested structure of the US biofuel mandate.

advanced biofuels") was mostly sugarcane ethanol from Brazil (although the automatic 1.5 overage factor for the biomass-based diesel mandate impinges on sugarcane ethanol's space). The situation in panel (a) depicted reality through 2010. However, in 2011, the situation changed and that requires us now to explain why the biodiesel model prediction error was high in the previous chapter.

Soaring Sugarcane Ethanol Prices in 2011 and Impacts on US Biodiesel Prices

For reasons explained in the following chapter, Brazil's ethanol prices went as high as six Brazilian Reals per liter in June of 2011 and, therefore, sugarcane ethanol prices itself out of the advanced mandate and supplies were not forthcoming. As shown in Figure 9.2 in Chapter 9, ethanol imports plummeted in 2011 and biodiesel net imports came close to zero. This put unexpected pressure on domestically produced biomass-based diesel consumption to fill the advanced mandate. Biodiesel prices went up relative to diesel prices in 2011 but could not bid up the price of soybean oil up fast enough to eliminate excess profits in US biodiesel production for that year. Therefore, production capacity constraints that could not consistently keep profits close to zero explain the prediction error in Figure 9.3 of the previous chapter. The source of this unexpected and sharp increase in biodiesel demand was due to the nested structure of the US mandate where no other biofuel was allowed to meet the advanced mandate.

Panel (b) of Figure 10.1 puts this in perspective for 2011 (and again in 2012 although there were no excess profits in US biodiesel production that year). The biomass-based diesel mandate was exceeded (mostly biodiesel) and that, along with the automatic 1.5 equivalence factor, results in a much lower residual for the advanced mandate. In the key months of 2011, sugarcane-ethanol from Brazil was not forthcoming; therefore, biodiesel filled the gap, but not without consequences for biodiesel prices and profits in biodiesel production (and, thus, prediction error in our model).

The prediction error for soybean oil in 2013 depicted in Figure 9.3 of the previous chapter is due to the blend wall (where regular cars are unable to consume more than 10 percent ethanol in the fuel mix and, therefore, biodiesel was unexpectedly called upon to overcome the blend wall and so capacity constraints and excess profits ensued). However, we first have to understand blend mandates and that requires an understanding of how the US mandate is implemented. To that we now turn.

The US Blend Biofuel Mandate

Although the US blend mandate was originally described and formalized in de Gorter and Just (2007b, 2009b), the profession has held out that the mandate is implemented as a consumption mandate because the EPA specifies each year a fixed quantity for each category of biofuel specific in

Table 10.1 and Figure 10.1. It is true that a fixed quantity of each biofuel is specified annually but the EPA implements the RFS as a blend mandate for each of the two mother fuels: gasoline and diesel. The causes of confusion by economists are described in detail in Box 10.3 which makes clear that in the end, there is a blend mandate for "motor gasoline" (the blend of ethanol and gasoline) and "motor diesel" (the blend of diesel and biomass-based diesel—biodiesel and renewable diesel).

After going through all of the complex macerations as described in Box 10.3, one can come up with mandated versus actual blend mandates for each year in the United States. We give these in Figure 10.2. This figure has never been calculated by others so this is the first time the reader has come across the graph, even though it should be one of the most distributed graphs on ethanol. But to be fair, some have tried. For example, Serra and Zilberman (2013, 141) state that the blend ratio was 5.5 percent in the year 2009, *"well below the US blend wall of 10 percent of ethanol in gasoline."* However, Figure 10.2 shows the blend ratio actually was 8.00 percent in 2009. That is actually a fairly big difference and will wreak all kinds of havoc on your economic analysis of biofuel policy if one does not understand how to calculate a blend ratio.

Box 10.3 What is a flexible-fuel vehicle?

A flexible fuel vehicle (often referred to as an FFV), or a flex car for short, is an automobile that can take a higher share of ethanol in the fuel mix with gasoline relative to regular vehicles. Regular cars can take a maximum of only 10-percent ethanol [25 percent for modified (non-flex) regular cars in Brazil] to avoid potential damage to their car engine. E15 has been approved in the United States but motorists are unwilling to have their cars consume more than 10-percent ethanol.

Flex-cars are widely used, especially in Brazil where currently 23 percent of cars are flex vehicles, but the number is growing rapidly as over 80 percent of new car sales in recent years are flex. A typical Brazilian flex car can run on a fuel with the share of ethanol between zero and 100 percent (E100); this range is zero to 85 percent for flex cars in the United States and European Union. In 2011, there were 3,266 (3,521) gasoline stations in the United States (European Union) concentrated in the Midwest (Sweden and France) selling E85. There are no "flex cars" or trucks for biodiesel although 20 percent is often cited as the maximum a regular car or truck can easily absorb (except for in very cold temperatures in the winter when the share has to be below 5 percent).

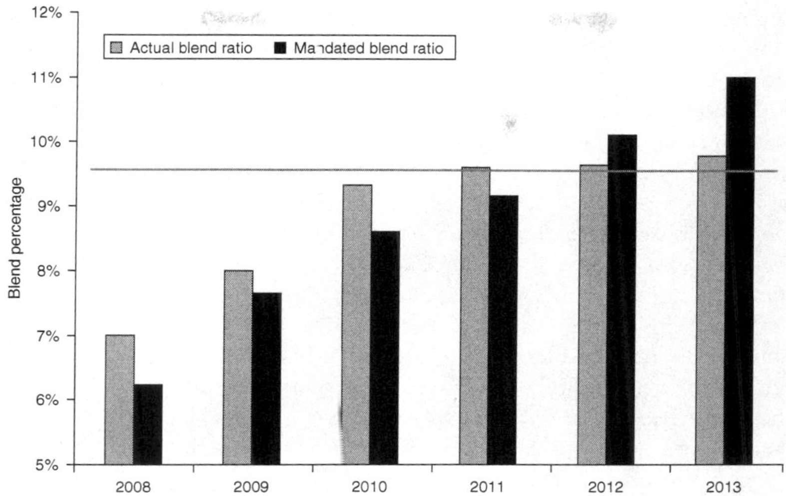

Figure 10.2 Actual versus mandated ethanol blend ratios.

Source: Energy Independence and Security Act of 2007 (Public Law 110–140), EPA (2014).

The information in Figure 10.2 is important for several reasons.

First, that is the only way to model the economics of a blend mandate and such data is required as an input (see the classic paper by de Gorter and Just, 2009b). Throughout the book (for each country), we transform the ratio E/G (B/D) into E/(E+G) [B/(B+D)], where E is ethanol, G is gasoline, B is biomass-based diesel, and D is diesel. This necessarily means that the transformed value of the mandate is lower than the announced value. The latter ratio is required to do a proper economic analysis of blend mandates.

Second, the data in Figure 10.2 are the only way to determine empirically what the blend wall actually is. The "blend wall" has to be defined in the same way the blend mandate works: regular cars are unable to consume more than 10 percent of the motor gasoline as ethanol (see Box 10.4). Papers associated with Wally Tyner at Purdue University argued that the blend wall was really 9 percent because Alaska and small blenders are exempt, along with several other factors (e.g., Tyner and Viteri 2010). Tyner and his colleagues argued the blend wall became an issue in 2010 although data shows that only in 2011 did the mandated blend ratio exceed 9 percent. However, then Tyner (2013) concedes the blend wall is 10 percent where he announces the "blend wall is about 13.3 BG, and that includes sugarcane ethanol." The Energy Information Administration (EIA) forecast for total motor gasoline for 2013 was 132.28 billion gallons. However, data in

Figure 10.2 would suggest the blend wall was approximately 9.5 percent (slightly lower than the maximum of the actual blend ratio because some ethanol is consumed as E85 fuel). Notice how the actual blend ratio is fairly constant from 2010 to 2013. This would imply the blend wall had been reached in 2010 already (and the ability to "water down the scotch" had been maxed out!). Therefore, why the blend wall was not 9 percent as Tyner originally argued it to be the case? The reason is that states and companies that were exempt still blended ethanol like the state of Alaska where over 8 percent of their motor gasoline contains ethanol. States may have had their own incentives or companies just wished to take advantage of "watering down the scotch." The point is one can make a strong case empirically that the blend wall seems to be 9.5 percent, not 9 percent or 10 percent, and that it has been reached for four years now. This is important information that we will exploit below.

A third reason why the data in Figure 10.2 is important is that one has to explain the gaps between actual and mandated ratios: positive gaps for the first four years but negative gaps in the last two years, 2012 and 2013. There are many reasons why companies would have exceeded mandated amounts in the earlier years, including the benefits from "watering down the scotch" and being able to store RINs for future use. The last two years have seen actual blend ratios remain constant but mandated ratios climbing. How was the gap filled? Taking 2013 as an example, approximately half of the gap was closed with stored RINs and the other with biomass-based diesel imports (recall Figure 9.2 in the previous chapter where the United States switched from being an exporter of biomass-based diesel to a net importer and vice versa for ethanol; the United States now is a net exporter of ethanol). The extra ethanol blended as E85 in these years is next to zero. This means the ethanol blend wall in the United States is being breached by imports of biodiesel (200 million ethanol equivalent gallons) and renewable diesel (300 million ethanol equivalent gallons).

The intuition or graphical summary for what happened in the years 2013 and 2014 is given in panel (c) of Figure 10.1. Biomass-based diesel production exceeds the advanced mandate, and with overage plus the 1.5 to 1.7 ethanol equivalence values, means there is no other advanced biofuel. Meanwhile, biomass-based diesel has also captured a share of the total RFS residual or renewable fuel gap.[9] This pushes corn-ethanol out [sugarcane ethanol has an implicit mandate with the California Low Carbon Fuel Standard (LCFS) but only a quarter of ethanol imports from Brazil in recent years goes to the LCFS] and, therefore, it is not surprising that US net ethanol exports turned positive in these two years.

The mandated amounts in the original Energy Independence and Security Act (EISA) of 2007 legislation for 2013 are compared to the actual amounts for each category of biofuel in Table 10.2. The actual cellulosic mandate was next to zero in 2013, while the biomass-based diesel mandate was increased by 280 million "wet" (actual) gallons; the mandate in 2007 was 1.5 billion ethanol gallons equivalent but the actual amount was 2.68 billion ethanol equivalent (see line for biomass-based diesel under advanced biofuel mandate). However, the advanced mandate was held at the original 2007 levels. This means waiving the cellulosic mandate in 2013 had no impact on total mandated levels; it simply means the mandated volumes of other biofuels were higher than the original 2007 legislated volumes. This has implications for biofuel and hence crop prices in 2013

Table 10.2 Mandates versus outcomes for 2013 (billion ethanol equivalent gallons)

	Mandated (EISA 2007)	*Actual*
Cellulosic Biofuels Mandate	1	0.014
Biomass-based Diesel Mandate[a]	1	1.28
Advanced Biofuel Mandate[b]	2.75	2.75
Cellulosic biofuel	1	0.014
Biomass-based diesel (1.5–1.7×)	1.5	2.68
Other advanced biofuel[c]	0.25	0.777
Total Advanced Overage	0	0.707
Renewable Fuel Gap[d]	13.8	13.8
Advanced overage	0	0.707
Corn-ethanol[e]	13.8	13.08
Total RFS Mandate[f]	16.55[f]	16.537[g]
Ethanol		13.55
Biodiesel		3.06

Notes:

[a] The mandated amount of biomass-based diesel is in actual or "wet" gallons. Biomass-based diesel counts 1.5–1.7 times actual gallons toward the remainder of the mandates.

[b] Advanced biofuels is renewable fuel (other than ethanol derived from corn starch) that is derived from renewable biomass, and achieves a minimum 50-percent reduction of greenhouse gas (GHG) emissions relative to gasoline or diesel it is assumed to replace.

[c] This includes imported sugarcane ethanol, bio-butanol, renewable diesel, and other advanced biofuels. For 2013, Brazilian sugarcane ethanol was 0.457 billion gallons and most of the remainder (0.32 billion gallons) is imported renewable diesel.

[d] The renewable fuel gap is the difference between the total renewable fuel standard (RFS) and the advanced biofuel mandate.

[e] There is no corn-ethanol mandate. Corn-ethanol can only fill the residual of the renewable fuel gap which in theory can be filled by any advanced biofuel. However, up to 2013, corn-ethanol has been the only biofuel counted toward the renewable fuel gap except for minute amounts of biodiesel, for example.

[f] Renewable fuel is any biofuel blended that achieves a minimum GHG reduction of 20 percent.

[g] Mandated.

[h] Actual.

Source: Energy Independence and Security Act of 2007 (Public Law 110–140), EPA (2013, 2014), Drabik (2011)

and especially for later years if the EPA continued this "policy," which it did not for 2014. This has caused political furor but, ironically, ethanol and biodiesel producers essentially are lobbying for higher mandates for their particular fuel beyond that legislated. Clearly, as cellulosic mandates increase exponentially, and cellulosic mandates are continually waived, then the pressure on biofuel prices and hence corn and soybean prices would mount. Apparently the EPA was aware of this and broke "tradition" by scaling back the 2014 mandate (although the final decision has not been made at the time of this writing).

The total advanced "overage" was 707 million ethanol equivalent gallons in 2013 that spills into the residual of the total RFS (the renewable fuel gap) and displaces not only corn-ethanol but also Brazilian sugarcane ethanol (an "advanced biofuel"). Corn-ethanol counted 13.08 billion gallons to the RFS in 2013 (720 million less than that which was specified as "corn ethanol" in the original 2007 EISA legislation). However, it is important to understand that there is no "corn-ethanol" mandate. Corn-ethanol can only fill the residual of the renewable fuel gap, which, in theory, can be filled by any advanced biofuel. However, up to 2013, corn-ethanol has been the only biofuel counted toward the renewable fuel gap except for minute amounts of biodiesel, for example, that did not meet the 50-percent reduction in the GHG emissions threshold.

Table 10.2 also shows that the mandated amount of biofuel (16.55 billion gallons) in 2013 is almost equal to the number of RINs generated for biofuels (16.537 billion ethanol equivalent gallons). The number of RINs is not the same as the actual consumption of biofuels (15.3 ethanol gallons equivalent for 2013) due to accounting errors, spills, and inexact data that will be revised. Notice that the actual wet gallons of biofuels mandated in 2007 for 2013 is unknown because the EPA does not tell us how much of each mandate category is ethanol versus biomass-based diesel, but it is obviously much lower than was actually consumed in 2013 because of the equivalence values of greater than one that were applied.

Furthermore, in 2014, the EPA reduced the total mandate amount by 1.34 billion gallons, equivalent to a decline of 0.89 billion "wet" gallons of biodiesel. This represents the entire US biodiesel production in 2012! Therefore, one can expect the impact of biofuel policies in 2014 to be much less in terms of its impact on corn and soybean prices.

The Economics of the Ethanol Blend Wall

The "blend wall" simply means regular cars are limited to a 10-percent blend of ethanol; otherwise, there is a potential damage to the engine. Flex cars can consume more than 10-percent ethanol (see Box 10.4). As

Tyner has correctly argued over the years, the blend wall is really less than 10 percent because of exemptions, including seasons due to cold weather, for example. We showed the United States reached a blend wall of approximately 9.5 percent starting in 2010 in Figure 10.2 and it has stayed fairly constant since then.

Box 10.4 Calculating the US blend mandate ratio: why economists are confused

Most economists model the US biofuel mandate as a fixed quantity for each of ethanol and biodiesel. Although a fixed quantity is specified in the EISA (2007) legislation, the actual way the EPA implements the RFS is very unique: in November of each year, it specifies a required ratio of each category of biofuel (cellulosic, biomass-based diesel, and advanced, plus the total) to the total forecast consumption of motor gasoline and motor diesel in the following year furnished by the Energy Information Agency. Because gasoline and diesel forecasts are not separated out, and because each category of biofuel includes both ethanol and biomass-based diesel, it is extremely difficult to calculate the amount of ethanol (biodiesel) mandated as a ratio of gasoline (diesel). The official forecasts are buried in a technical document with confusing variable names, in the middle of other variable names that are not germane to the exercise. Determining the amounts of ethanol and biodiesel in each category of biofuel is also a very complicated task requiring painstaking work through the complex formulas published in the RFS "Final Rule" (EPA 2013b), with adjustments for exemptions that vary substantially each year, some of which are not given because of confidentiality. Once the data has been located, it may seem straightforward to divide the EIAs forecast of renewable gasoline (mainly ethanol) by motor gasoline as Tyner does. This is incorrect when modeling the RFS because the mandates specified by the EPA are significantly larger than the EIA forecast, meaning more biofuels must be used in order for blenders to meet their blend compliance.[10]

Therefore, the confusion for economists is perpetuated by the EPA and EIA by not giving the data that requires one to calculate the blend mandate ratio for ethanol (as shown in Figure 10.2) or biodiesel. These blend ratios are required to properly model the economics of a blend mandate as specified in de Gorter and Just (2009b). However, almost all economists are not even aware that they have to calculate a blend mandate ratio. How can that be? Well, it appears to be a case of Say's law: "supply creates its own demand." Economists are furnished (in Table 1 of EPA 2013a) with the official policy that a specified amount of cellulosic biofuels, biomass-based biodiesel, advanced biofuels of ethanol, and total biofuels has to be consumed each year. Therefore, it must be so. They are then furnished with (economically) meaningless "percentage standards" in Table 2 of EPA (2013a) that cannot be used by economists for analyzing either consumption or blend mandates. Apparently, such ratios are required for the EPA to ensure each of the nested mandates are fulfilled and so need not concern economists.

It is almost as if the EPA and EIA intentionally mislead economists or did not understand the economics of the US RFS in the first place. The latter is certainly correct as we have phoned many people in each agency. However, a quick perusal of the Web suffices:

> Peter Gross, a renewable fuels expert at the US Energy Information Administration (EIA) explained to me how RINs work in response to an email query: All obligated parties (refiners and importers of refined fuel products) must satisfy their "renewable volume obligation" (RVO) which is essentially their share (based on how much fuel they produce or import) of the total renewable fuel that must be used [RR: For 2013 the proposed volume is 16.55 billion gallons][11]

This is very misleading, if not patently false. It clearly states that the RFS is a consumption mandate; "obligated parties" must fulfill their share of the Required Volume Obligation (RVO), which is then defined as the total mandate of 16.55 billion gallons. However, that is not how it works.

The correct way to explain it is to first show how one calculates a blend mandate ratio (i.e., mandated ethanol divided by mandated ethanol plus forecast gasoline consumption). This, along with the analogous biodiesel mandate ratio will generate the total volume of biofuels consumed to equal the total mandated volume, provided the actual consumption of each of gasoline and diesel are exactly equal to their forecast values by the EIA in November of the previous year. If actual gasoline consumption turns out to be half of the forecast value, and "obligated parties" only blend the minimum required quantities of ethanol, then total ethanol consumption in that year will be half the mandated volumes. Furthermore, no adjustment for this is made in the next year.[12] This shows the US biofuel mandate is a blend mandate. The ability to bank and borrow "obligations" in the form of RINs complicates things but is explained elsewhere in this chapter.

The EPA uses the term RVO everywhere in their documents and gives the false impression that the RFS is a consumption mandate. It is no wonder economists get confused, and this leads to errors in analyzing biofuel polices in many dimensions (environmental, welfare, and interaction effects, just to name a few). It also leads to errors in understanding RIN prices (see corresponding discussion in text).

The total amount of ethanol (in wet gallons) that was forecast by the EPA to be blended in 2013 was 14.63 billion gallons to exactly comply with the total RFS, and we estimated the blend wall from Figure 10.1 to be 9.5 percent of total forecast motor gasoline, or the blend wall is estimated to be 12.62 billion gallons, which means an extra two billion gallons was required to be sold outside of E10 markets in order to fulfill the blend mandate. There are several ways this requirement can be met.

First, one can reduce or eliminate "neat," that is E0, gasoline sales (accounting for 12 percent of total fuel sales in 2011—see Foster et al. 2011) by providing discounts on ethanol, provided RIN prices are high enough to induce a change in consumer behavior. Second, increased E15 and E85 sales are possible as the EPA recently approved E15 for cars manufactured after 2001 but the market has not grown very much because of higher supply costs and consumer resistance. E85 sales are also limited because of the high costs of E85 stations and the requirement for consumers to have flex cars (that can take almost any blend of ethanol and gasoline). However, sufficiently high RIN prices would induce an expanding E15 and E85 market by providing fuel price discounts to E15 and E85 prices on a miles per gallon basis. Third, an exogenous increase in fuel consumption (through lower oil prices or higher demand for miles traveled) can relieve the pressure on RIN prices as the required blend for 2014 and beyond may be lower than the current 11.02 percent, depending on forecast fuel consumption upon which each year's blend ratio is calculated upon. Fourth, to increase the use of drop-in biobutanol which allows conventional cars to consume E13 (instead of E10); moreover, every gallon of biobutanol counts toward the mandate as 1.3 gallons of corn-ethanol.[13]

A fifth option is that approximately 2.5 billion gallons worth of RINs had been carried forward, which is more than enough to eliminate the excess ethanol that needs to be sold. However, the blend wall in the following year (2014) may be an even greater constraint; therefore, only those RINs that expire in 2013 likely were used (stored RINs have a 24 month shelf life), which still would have been enough, but RIN prices in 2013 indicated otherwise. Finally, every gallon of biodiesel (renewable diesel) blended counts as 1.5 (or 1.7) gallons toward the total RFS mandate. Therefore, biodiesel sales above the biodiesel mandate count toward fulfilling the ethanol mandate.

The last two options—using stored RINs and biomass-based diesel spilling over into the renewable fuel gap—is what really happened in 2013 and now in 2014 [as we showed in discussing Table 10.2 and panel (c) of Figure 10.1 earlier]. This caused D6 RIN prices to skyrocket in 2013 to over $1 per gallon and have recently been approximately 40 cents per gallon, well above historical levels. An explanation of this is given in the next section.

To summarize, the blend wall issue has drawn a lot of attention recently amongst the fuel industry, media, politicians, and pundits on the economics of biofuel policy. We think the most efficient way to deal with the blend wall is to continue with a RIN market. Higher D6 RIN prices means the consumer of fuel (10-percent ethanol, i.e., E10) will pay the cost to have ethanol consumed above the 10-percent limit on normal cars in the form of E85 consumption by flex cars. This also implies that the fuel blenders who blend beyond the 10-percent ethanol blend wall will be implicitly subsidized to reduce consumer prices of ethanol for E85 sales to absorb the required excess of ethanol beyond the 10-percent "wall."

We do not think the "blend wall" is a problem. It just means the amount of mandated ethanol cannot be supplied to the market unless there is a government policy, in this case, a blend mandate. If D6 RIN prices had averaged $1 per gallon in 2013, then there will be a transfer from fuel consumers of approximately $1.26 billion to suppliers to induce less E0 sales and more E15, biobutanol (E13), biomass-based diesel beyond its mandate and E85 sales to absorb the 2 billion ethanol gallons equivalent. Given that E10 consumption is forecast to be approximately 135 billion gallons, and with an average E10 price of about $3.50 per gallon, a $1 per gallon RIN price translates into an implicit tax on $472.5 billion worth of E10, or an implicit tax of 0.27 percent. This should be lower as part of the required 1.97 billion gallons was met by using stored RINs.[14]

Some economists think we should subsidize E85 stations and flex cars but these do not exhibit public good characteristics. Furthermore, subsidies would be inefficient as they would not allow the market to decide which of the various avenues (and by how much) we described earlier would be applied to ensure two billion gallons of ethanol beyond the blend wall are consumed.

Mandate Compliance Through the Use of RINs

Prior to 2013, the D6 RIN price mostly reflected the costs of having to sell ethanol outside the E10 market and the formula in Figure 10.3 worked fairly well (where D6 RIN prices followed the difference between gasoline and ethanol prices).[15]

However, this D6 RIN price formula fell apart in 2013 as biomass-based diesel overage into the renewable fuel gap caused all RIN prices to equalize. The value of D4 (biomass-based diesel), D5 (advanced biofuel including sugarcane ethanol) and D6 (corn-ethanol) RIN prices over time is given in Figure 10.4.[16] In 2013, all RIN prices converged.[17] The D4 RIN prices followed the same formula (divided by 1.5 as D4 RIN prices are in $ per ethanol gallon equivalent) and continue to date as it determines the marginal value of all RIN prices.This is an important point. One must understand that D4 RIN prices limit the market for D6 prices as a blender would choose the cheapest compliance option at the margin, meaning that the biodiesel blend margin partly determines the E85 price relative to E10.

D6 RIN prices reflect some blenders' willingness to pay other blenders to absorb that ethanol above the 10-percent blend wall by having them exceed their 11.02-percent minimum required blend in 2013. It should be noted that both D6 RIN prices and the E85 price discount relative to E10 rose sharply and in tandem in the first half of 2013 and followed each other down in the second half of 2013

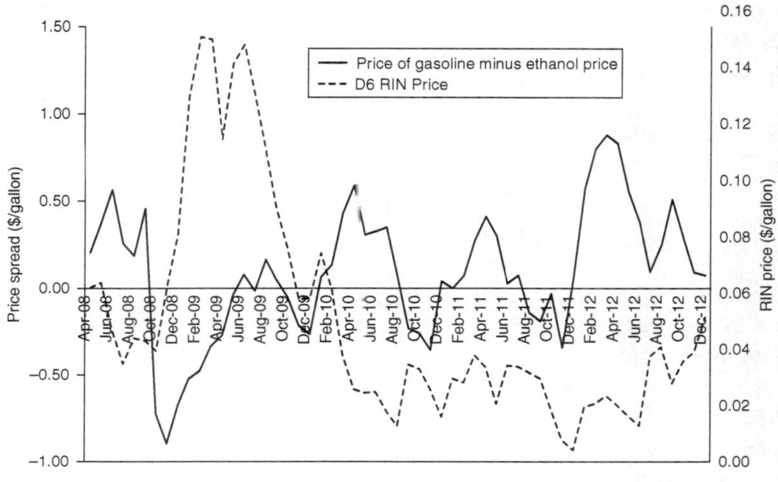

Figure 10.3 D6 RIN prices versus price of gasoline minus ethanol price.
Source: OPIS, CARD (2014a), EIA (2014a).

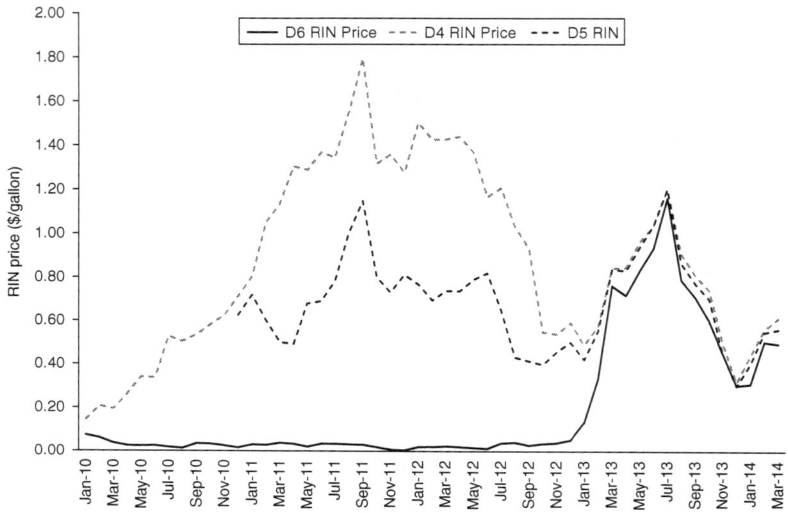

Figure 10.4 Historical RIN prices.
Source: OPIS.

D6 RIN prices did not apply to all corn-ethanol consumption but relate only to the estimated two billion gallons that have to be sold outside of E10. Therefore, RIN prices do not act as a subsidy to ethanol producers and are likely to have minimal effects on ethanol market prices. Because RIN prices reflect the increased cost of RFS compliance and so E10 consumer prices rise to pay for increased E85 sales, if anything, the ethanol price declines. However, if the blend mandate is adjusted the following year, then the RFS will be achieved, thus resulting in no change in the ethanol price in the long run.

What RIN Prices Are Not

Economists are continually confused about the economics of RINs but the three most important mistakes are that a RIN price is (1) the mandate premium over the tax credit and thereby reflects the extent to which the mandate is binding (e.g., McPhail et al. 2011), (2) a subsidy to ethanol (and hence corn) producers (e.g., Holland et al. 2014), and (3) is the difference between the supply and demand price, and applies to all consumption (e.g., Babcock, 2012c), and, therefore, is analogous to a production quota or the price of a cap and trade permit.

The RIN price does not determine the extent to which the mandate is binding; the RIN price is a result of policy and market forces and merely represents the compliance costs and, therefore, has not impacted the levels and volatility of corn prices. In fact, the opposite is true when the blend wall is an issue. If oil prices decline or fuel demand shifts out, the blend mandate becomes more binding (defined as the mandate premium over the tax credit) whereas the blend wall constraints become relaxed and so RIN prices decline (further reducing E10 prices). On the other hand, if corn supply conditions deteriorate and shift up the ethanol supply curve, RIN prices increase and the blend mandate becomes more binding. We showed in Chapter 2 how to determine the mandate premium over a tax credit and it does not involve RIN prices. D6 RIN prices for corn-ethanol have been a few pennies per gallon over most of the time period until 2013, whereas we showed in Figure 3.2 that the mandate premium has been very high at times in the past.

RIN prices do not act as a subsidy to ethanol producers because the ethanol price (and, therefore, position on the ethanol supply curve) is determined by the mandate. In a simple world of only corn-ethanol, RIN prices simply subsidize E85 consumption and are a tax on E10 consumption (de Gorter et al. 2013e).

The reason why RINs are not analogous to a production quota or a cap and trade permit is because the ethanol blend mandate is a *minimum* percent requirement. If one blends less than the minimum, you do not free up a RIN (unlike with a production quota or cap and trade permit). RINs are forfeited or "retired" to the EPA—they are not kept and sold on the free market. A RIN is freed up only if you blend *above* the minimum required. Abstracting from speculative trades, a RIN is sold to those firms who blend *less* than the minimum (otherwise, the RIN has no value). RINs have value only for the ethanol blended beyond the required minimum (for a production quota or cap and trade permits, their value is on the total production or permitted pollution). Of course, the RIN can be stored and sold later, but the same concept holds; it is just that the buyer of the RIN is blending less than the minimum required but in some other time period. Therefore, the economics of RINs are very different and require a thorough understanding of how a blend mandate system works and of the economics of blend mandates as developed in this book.

Explaining the Prediction Error for Soybean Oil in 2013

We are now in a position to explain our model's soybean (rapeseed) oil-price prediction errors in 2013, depicted in Figure 9.3 of the previous chapter. The price of RINs for biodiesel (the biodiesel RIN price reflects the cost of complying with the biodiesel mandate) increased and converged with that of RIN prices for sugarcane- and corn-ethanol. The incentive to increase biodiesel production was great as biodiesel and renewable diesel were called upon on short notice to not only fill the entire mandate for advanced biofuels but also to contribute to the residual component of the US mandate where corn-ethanol is eligible. Furthermore, biodiesel production went up significantly, as witnessed by the sharp increase in the number of biodiesel RINs generated. However, production did not increase fast enough in 2013 and blenders turned to huge amounts of imports of biomass-based diesel[18] to fill their requirement, meaning they did not have to bid up domestic soybean as high as predicted. By the end of 2013, profits returned to normal (and both D4 RIN and biodiesel prices declined).

Concluding Remarks

The importance of the nested structure of the US RFS has impacted the ethanol and biodiesel, and by extension, corn and soybean prices. It has also impacted biofuel trade flows, excess profits in ethanol and biodiesel production, and, therefore, price volatility. We derived some very important results in this chapter. We showed the US mandate is a blend mandate and the appropriate blend ratio was calculated, showing the blend wall has been "binding" since 2010. We outlined a basic economic model of the blend wall and empirically found it was not being overcome by expanded E85 sales but by biodiesel and renewable diesel, imports of which have exploded in recent years. Meanwhile, corn-ethanol exports have increased sharply. The decision to reduce the mandate for 2014 has a big impact on corn and soybean prices.

We showed that RIN prices are not a subsidy to ethanol production nor do they reflect the extent to which the mandate is binding. Furthermore, RINs do not act like production quotas or cap and trade permits. It is very unique, along with the economics of mandates and tax credit, a careful formulation of which is required in order to grasp the significance of the complex structure of the US RFS. This chapter highlights the importance of understanding the basic economics of biofuel policies; otherwise deciphering the economic impacts of the complex interaction effects between the various sub-mandates within the US RFS will be misunderstood and lead the analyst astray.

CHAPTER 11

THE ECONOMICS OF BRAZIL'S SUGARCANE–ETHANOL/SUGAR COMPLEX AND POLICIES*

The previous chapter emphasized the importance of the complex structure of the US mandate and how the nested mandates interact with one another. Brazilian sugarcane ethanol is central to the outcome, being an advanced biofuel. Brazil was the world's largest ethanol producer until the United States surpassed it in the mid-2000s, but Brazil is still the world's largest net exporter of ethanol. However, this may not last either as our discussion in the previous chapter showed how the ethanol blend wall in the United States has made biomass-based diesel the biofuel of choice (especially because of the 1.5 to 1.7 ethanol equivalence values) and has converted the United States into a significant net exporter of ethanol in recent years. Brazilian sugarcane-ethanol, in 2013 and 2014, has to compete head on with corn-ethanol (before, it received higher D5 RIN prices as an advanced biofuel). The only advantage Brazil's sugarcane ethanol receives is higher prices with the California Low Carbon Fuel Standard (LCFS); thus, there is still two-way trade: Brazil exports sugarcane-ethanol to the United States and the United States exports ethanol to Brazil. These types of ethanol are identical products; it is just that their greenhouse gas (GHG) emissions thresholds differ and, therefore, the way in which binary GHG emission thresholds operate in the US renewable fuel standard (RFS along with the California LCFS) causes the two-way trade (Meyer et al. 2012).

It is also important to analyze Brazilian ethanol markets and policies because world trade links world ethanol prices. Therefore, Brazil's ethanol and gasoline policies since 2010 directly impact US ethanol (and, thus, corn) prices, as do shocks in the sugar market (de Gorter et al. 2013f). Moreover, given the law of one relative international price of grains and oilseed prices, a change in Brazilian ethanol prices have an impact on all grain and oilseed prices.

Sugarcane in Brazil is processed into sugar and/or ethanol, often in flex plants that can switch between the two products, depending on prices. We summarize an economic model of flex plants in this chapter, with export demands and two domestic fuel demand curves: E25, a 25-percent blend of ethanol with gasoline consumed by conventional cars, and E100, ethanol consumed only by flex cars. Flex cars can switch back and forth from E100 to E25, depending on relative prices, but regular cars cannot. Therefore, a unique switching function has to be estimated. This makes the total demand for ethanol in Brazil more elastic compared to a situation with no flex cars. Our economic model allows us to analyze the market impacts of specific Brazilian policies that are very controversial, namely varying the E25 blend mandate, fixing market gasoline prices below world prices, altering the high gasoline tax, and having a higher tax exemption for ethanol used in E25 than in E100.

Because of two demand curves, with flex car owners switching between fuels depending on relative prices, and because the mandate is for E25 only, the impact of each Brazilian policy in theory has an ambiguous impact on ethanol and sugar prices. Conventional wisdom is that a higher level of the mandate, gasoline tax exemptions for ethanol and gasoline price, and a lower gasoline tax, all help the ethanol industry. But for two policies—a low gasoline tax and a high tax exemption for ethanol used in E25—our empirical results show ethanol and sugar prices decline.

Brazil ethanol was touted as a success but the United States corn-ethanol production overtook it and is now double the size. The US allowed the ethanol import tariff to expire at the end of 2011, yet Brazilian exports to the United States have stagnated and are mostly due to advantages over corn-ethanol because of the binary sustainability standards of the RFS and California's LCFS. We discuss the high-cost structure of Brazilian ethanol production later in the chapter.

Therefore, because of the uniqueness in both the production and demand structure for ethanol in Brazil, the transmission of an ethanol price shock to the sugar price is much lower than a corresponding response in the corn price. Empirically, we estimate that a \$0.1 per gallon increase in the ethanol price in Brazil in 2011 would increase sugar prices by 1.8 percent, which is lower than the estimated 5.8-percent increase in corn prices but higher than the estimated 1-percent increase in soybean prices presented in Chapter 9 for the same biodiesel price increase.

This chapter is outlined as follows: in the next section, we describe the basics of the Brazilian markets while the section after that outlines the formal economic model. Then, we explain the policy issues facing the Brazilian ethanol market. Before the conclusions, we illustrate how US ethanol policy developments had a big impact on Brazilian ethanol prices.

The Structure of the Brazilian Sugarcane–Sugar-Ethanol Industry

Brazil has developed a unique system of producing competing tradable products—sugar and ethanol—from non-traded sugarcane. State-of-the-art "flex-plants" can produce either sugar or ethanol from sugarcane, and within a production year, can switch between the two products such that each product uses between 35 percent and 65 percent of the sugarcane. Furthermore, under ideal conditions, flex-plants can extract up to 18.6 liters of ethanol per metric ton of sugarcane processed into sugar from molasses, a by-product of sugar production. The ethanol produced in Brazil has two forms: anhydrous and hydrous. Anhydrous ethanol cannot be consumed in its pure form and is, therefore, mixed with gasoline. Any car in Brazil is able to take a typical the blend (up to 25 percent) of anhydrous ethanol and gasoline. On the other hand, hydrous ethanol is only consumed in its pure form by what is also called in Brazil "flex-cars"; at the same time, these special cars can also take any share of anhydrous ethanol. This leads to two different demand curves for ethanol as transportation fuel: an *anhydrous* ethanol–gasoline fuel mixture E25 and E100 (100-percent *hydrous* ethanol).

The share of anhydrous ethanol in total ethanol consumption has increased from 33 percent in 2009–2010 to 45 percent in 2011–2012. There has been a narrowing of the "parity gap" between fuel and E100 prices. In other words, the price of E100 is usually discounted to the price of the blend of anhydrous ethanol and gasoline in terms of the cost per kilometer traveled in order to overcome the Brazilian blend wall (of 25 percent). As shown in Figure 11.1, from 2003 to 2011, the price of E100 was at a substantial discount (often as high as 25 percent but averaged 20 percent from 2003 to 2010) compared to the fuel parity (cars get about 30 percent less kilometers per liter of E100 relative to the E25 fuel). This encouraged the adoption of flex cars and infrastructure like E100 dispensing stations. However, as the share of flex cars has increased, this discount has disappeared in recent years (Figure 11.1).

Brazil was regarded as the lowest cost producer of ethanol in the world; however, since mid-2009, Brazil's market price of ethanol has surpassed that in the United States and the two prices now track each other fairly closely. Brazil ethanol exports peaked at almost 20 percent of its production in the 2008–2009 crop year but exports were only 1.9 billion liters in 2010–2011 with imports of 1.3 billion liters, mostly from the United States.

The high ethanol price is partly due to the very high cost to expand sugarcane production in Brazil in the near term. This in turn is in part due to the very strong currency but also due to the new land being less

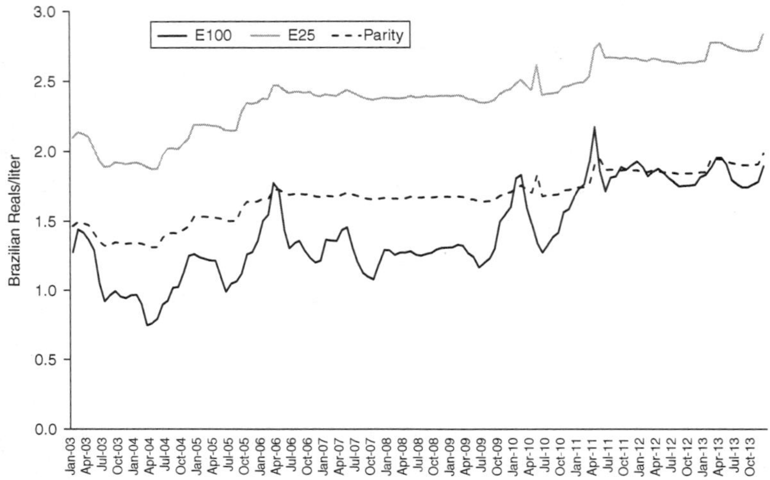

Figure 11.1 Fuel prices in Brazil (São Paulo).

Source: de Gorter et al. (2013f). Using CONAB data.

productive than the existing land in sugarcane production. Another key reason for high domestic prices of ethanol has been the revaluation of Brazil's currency; the Real has gained approximately 30 percent in value in past half decade relative to the US dollar. As the Real revalues, the price of ethanol in US dollars increases (or the US price in Reals declines). The strong currency is due to strong economic growth in Brazil propelled by booming commodity exports, especially to emerging economies such as China. Other factors contributing to high ethanol prices in Brazil include world record sugar prices, bad weather affecting the sugarcane harvest around the world, increasing demand for sugar, government policies for sugar in India and for ethanol in Brazil, and strong domestic demand for ethanol.

The strong demand for ethanol in Brazil is caused by growing incomes, causing car sales to boom and miles traveled per car to increase. Brazil has only one car for every five inhabitants whereas the ratio in the United States is essentially one-to-one. Recently, approximately 50 percent of total gasoline plus ethanol consumed in Brazil has been ethanol, compared to 10 percent in the United States. Brazil has significant flexibility, unlike in the United States, to increase domestic ethanol use because of the large infrastructure of flex fuel cars and dispensing pumps. To illustrate, gasoline consumption (excluding any anhydrous ethanol mixed with it) in Brazil increased by two billion liters from 2000–2001 to 2009–2010 but ethanol consumption increased a whopping 24 billion liters. However, since 2009, gasoline consumption increased 12.65 billion liters whereas ethanol consumption has declined almost 21 percent (see Figure 11.2) as rising ethanol prices, along with lower gasoline taxes and gasoline prices pegged below world prices (see Figure 11.3) all have resulted in a higher share of gasoline being consumed.

Ethanol consumption reached a peak of 50 percent of total fuel consumption in 2009–2010 but fell to 30 percent in 2013 (compared to 10 percent in the United States). In 2008, the United States consumed 18 times the amount of fuel (blend of gasoline and ethanol) per capita compared to Brazil. Therefore, if Brazil is going to catch up to the United States in total per capita fuel consumption, and supply and demand factors in sugarcane, sugar, and ethanol markets are such that ethanol prices continue to be high, it will be very difficult to maintain a 50-percent share of ethanol in gasoline-based fuel consumption, especially if Brazil's overall economy continues its high growth rate.

Although Brazilian ethanol prices have been at record levels in recent years, investment in Brazilian sugarcane-ethanol has stagnated since the 2008 financial crises, after having increased 10.4 percent per annum between 2000–2001 and 2008–2009 (de Gorter et al. 2013f). Furthermore, investments in new sugarcane-processing plants have almost come to a halt as only two new plants were built in 2012–2013 (and only four more are foreseen by 2015; see de Gorter et al. 2013f); this should be contrasted

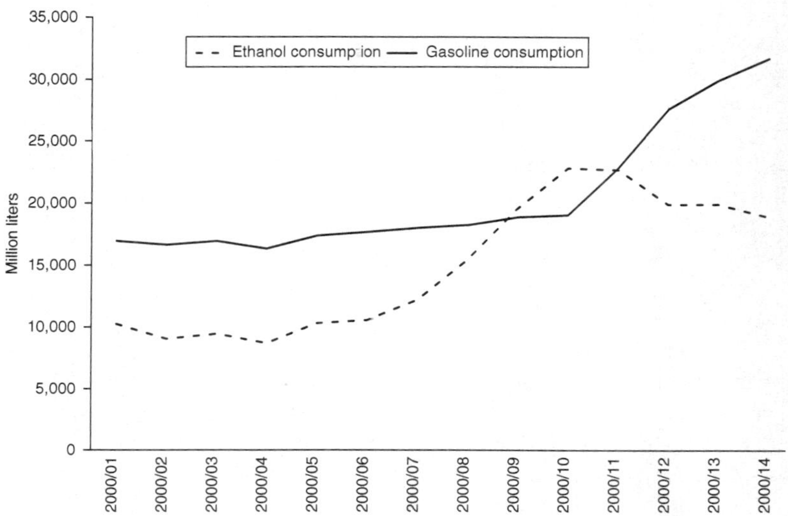

Figure 11.2 Gasoline versus ethanol consumption in Brazil (million liters).

Source: de Gorter et al. (2013f). Data: Brazil ANP, U.S. EIAR.

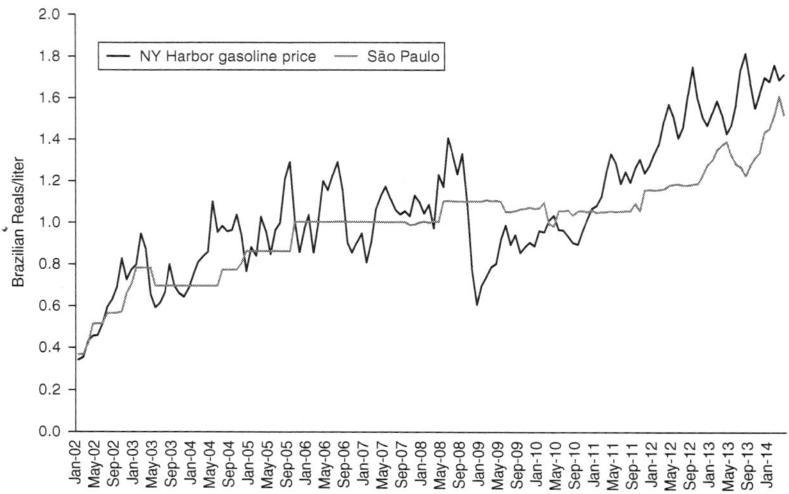

Figure 11.3 US and Brazil gasoline prices compared.
Source: de Gorter et al. (2013f) and ANP (2012).

with the 2008–2009 peak when 30 new production units were built. To fill projected ethanol demand, it is estimated that, on average, 12.5 plants need to be constructed annually between 2014 and 2020. In addition, in order to maintain world market share in sugar exports, given sugar's projected demand growth, sugarcane production will need to more than double, and 32 new sugarcane mills are predicted to be required by 2020.

Significant investment and time is required to have new production come online. Sugarcane is a perennial crop with multiyear harvests (ratoons) associated with one planting. Because of insufficient capital due to high interest rates to expand sugarcane production, the crop rotation has been expanded in Brazil, resulting in lower yields of sugar and ethanol per hectare sugarcane harvested. Private investment is also hampered by the lingering effects of the 2008 financial crisis and the lack of infrastructure to move ethanol to ports.

Meanwhile, world sugar prices in recent years have been at record levels, mainly because of growing worldwide demand, fostered by rising wealth and population in emerging economies, and bad weather affecting sugarcane production not only in Brazil but also in India—the world's second largest sugar producer. Government policies in India may have been a contributing factor to the run-up of sugar prices in recent years. Governments set a price support for sugarcane as well as imposing a sugar levy, by which they are required to sell at below market price to the public distribution centers. In addition, the government intervenes in limiting sugar exports and imposes marketing restrictions, such as limits on private stockholdings. India faced sharply lower sugar production in 2008 and 2009 and Australia's sugar crop was adversely affected by bad weather in 2010. Furthermore, bad weather reduced Brazil's sugarcane crop by approximately 10 percent in 2011–2012 and was expected to remain flat for the 2012–2013 marketing year (approximately 570 million tonnes). Brazil's sugar production is approximately one-quarter of the world total and Brazil exports about two-thirds of total world sugar trade.

An Economic Model of Modern Sugarcane Flex Plants

Although the yields of sugarcane products vary year by year (e.g., because of weather conditions), a metric ton of sugarcane produces approximately 130 kilograms of sugar or 70–75 liters of ethanol (depending on the type of ethanol, with anhydrous ethanol having a lower yield). Regardless of the use of sugarcane for sugar or ethanol, bagasse—a fibrous matter—is produced and is burned in special boilers to cogenerate electricity and steam. The electricity generated from bagasse contributes to the economics of sugar and ethanol production, not only by avoiding the purchase of electricity but also by selling the excess electricity on the grid.

Another distinguishing feature of the sugarcane processing in Brazil is that a by-product of sugar production, called molasses, is used to produce ethanol as it contains a lot of sugar that cannot be further extracted. This is the indirect channel through which ethanol is produced in Brazil. Because ethanol competes with sugar, Brazilian ethanol differs from the US corn-ethanol as the latter is the main product of corn processing, and unlike molasses, the dried distillers grains with solubles (DDGS) is returned to the corn market. Flex plants can potentially extract up to 18.6 liters of ethanol from molasses per metric ton of sugarcane processed into sugar. Therefore, if 55 percent of total sugarcane production was devoted to sugar production, and every plant maximized ethanol production from molasses, then 25 percent of total ethanol production in Brazil could come from molasses alone. However, the observed yield is significantly lower—approximately 9 liters as plants dedicated to just sugar production are unable to extract ethanol from molasses. These production relationships are schematically depicted in Figure 11.4.

The competition among sugar and ethanol producers creates links among prices of sugarcane and hydrous and anhydrous ethanol:

$$P_{SC} = \phi_S P_S + \delta_H P_H + \delta_A P_A + \Psi_S - \phi_S \xi, \tag{11.1}$$

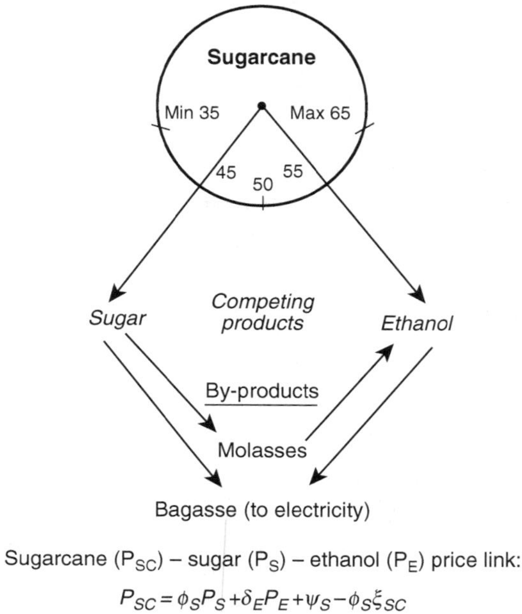

Figure 11.4 Economics of sugarcane–ethanol–sugar flex plant.

where P denotes a price and the subscripts SC, S, H, and A denote sugarcane, sugar, hydrous ethanol, and anhydrous ethanol, respectively. The remaining parameters characterize the yields of sugar, molasses, ethanol, and bagasse per metric ton of sugarcane (a detailed description of these parameters is presented in de Gorter et al. 2013f).

The prices of anhydrous and hydrous ethanol are positively linked on the supply side through the cost of sugarcane and processing costs

$$P_A - P_H = \varphi_0 + \varphi_1 P_{SC}. \tag{11.2}$$

This link has important implications because whenever the price of either ethanol type is affected (e.g., because the government changes a tax exemption), the price of the other ethanol type changes in the same direction.

The unique structure of sugar and sugarcane-ethanol production in Brazil invokes a question on how an effect of an exogenous shock in sugarcane-ethanol on the sugar price compares to the effects of a change in corn-ethanol and biodiesel prices on corn and soybean prices. Using the theoretical model and data in de Gorter et al. (2013f), we empirically estimate that a \$0.10 per gallon increase in the ethanol price in Brazil in

2011 would increase sugar prices by 1.8 percent, which is lower than the estimated 5.8-percent increase in corn prices for the same increase in ethanol prices (see Chapter 9).

Although there are two ethanol types in Brazil, only the consumption of anhydrous ethanol is mandated by requiring 18 percent to 25 percent of the volume of the final fuel, denoted as E25 in this chapter, to come from anhydrous ethanol. The Brazilian government changes the blend requirement depending, among other things, on the prevailing market conditions in a given year. The consumer fuel price is the weighted average of anhydrous ethanol and gasoline prices (inclusive of taxes), where the weights are the shares of ethanol and gasoline in the final fuel blend (m_F denotes a fixed marketing margin)

$$P_F = \alpha(P_A + t_A) + (1-\alpha)(P_G + t_G) + m_F. \tag{11.3}$$

The consumption of E100 is affected by the existing tax exemption as well as parity gap between the price of E100 and E25; the price of E100 is given by

$$P_{E100} = P_H + t_H + m_{E100}, \tag{11.4}$$

where t_H denotes the tax on hydrous ethanol and m_F denotes a fixed marketing margin.

Because of the Brazil's sugarcane-based ethanol fuel program (Pró-Álcool) that started in the 1970s of the last century, a dual vehicle structure has developed in Brazil. On the one hand, there are regular cars that can run on a mixture of anhydrous ethanol and gasoline with an upper (technological) bound on the share of ethanol. On the other hand, flex cars can take any share of ethanol either in its anhydrous or hydrous form. This implies an asymmetry in that flex cars can run on both E100 and E25, whereas the regular cars can only use E25. This has important implications as we show shortly. It also means that there are two different demand curves for ethanol as transportation fuel: an *anhydrous* ethanol–gasoline fuel mixture and E100 (100 percent *hydrous* ethanol).

A proper modeling of the Brazilian ethanol market has to take into consideration that the flex car owners are sensitive to the change in the price gap between fuel and E100. For example, when the price gap shrinks, many flex car owners will shift to fuel consumption. However, not all will do so, as documented by Salvo and Huse (2013) who analyze the responsiveness of Brazilian flex cars owners to the change in the relative price of fuel and E100. The authors find that roughly 20 percent of flex-fuel motorists choose fuel when it is priced 20 percent above E100 in energy-adjusted terms and, similarly, 20 percent of motorists choose E100 when it is priced

20 percent above fuel. Mathematically, we can express the consumption of the mixture of anhydrous ethanol and gasoline (D_F) as

$$D_F = f(P_F) - X(\boldsymbol{\theta}), \tag{11.5}$$

and the consumption of E100 (D_H) as

$$D_H = g(P_{E100}) + X(\boldsymbol{\theta}), \tag{11.6}$$

where f and g denote downward sloping demand functions for the fuel and E100 and the term $X(\boldsymbol{\theta})$ denotes a shifting factor determining the responsiveness of flex cars owners to a change in the relative price between fuel and E100.

It is important to note that if one converted the observed consumption of ethanol in Brazil into its sugarcane equivalent and then added to it the quantity of sugarcane processed into sugar, the resultant quantity would exceed the observed supply of sugarcane. This happens because of the additional ethanol extracted from molasses. Therefore, the market equilibrium requires a correction for the double counting of sugarcane for ethanol.

As the sugar and ethanol prices are closely linked through the competition for sugarcane and because Brazil is the largest sugar exporter (60 percent of the world total), the shocks to the world sugar market will affect not only Brazil's domestic price of sugar but also the price of ethanol. Because ethanol prices in Brazil and the United States are linked, policies in Brazil have repercussions for ethanol and corn prices in the United States. The same reasoning also holds for the changes in biofuel policies.

Policy Issues

The Brazilian government implements several policies that shape the sugarcane ethanol and fuel markets. First, Brazil has a mandate for anhydrous ethanol mixed with gasoline, requiring 18 percent to 25 percent of the total fuel mixture to be anhydrous ethanol, depending on supply–demand conditions. Second, E100 sales of hydrous ethanol enjoy an excise tax exemption greater than what is needed to compensate for the lower miles obtained from a gallon of E100 relative to a gallon of the gasoline–anhydrous ethanol fuel mixture. The tax on anhydrous ethanol is even lower, although consumers only see the blended fuel price. Third, the government has often in the past, and again recently, held the price of gasoline below world gasoline prices to avoid adverse effects on inflation. Fourth, the federal government has recently eliminated the gasoline tax.

de Gorter et al. (2013f) show that although each of these policies in theory has ambiguous market effects, under a weak assumption, which is almost always met in reality, most of the effects are unambiguous. Because it is important to understand where the ambiguity of the effects of Brazil's biofuel policies comes from, we illustrate it on the example of a reduction in the gasoline tax (Figure 11.5).

The instantaneous effect of a lower gasoline tax is a reduction in the price of fuel from P_{F0} to P_{F1} in Figure 11.5. As a result, the consumption of fuel increases by the distance *a*, 25 percent of which is anhydrous ethanol when the mandate is 25 percent (in other words, with E25). With this decline in the fuel price, the parity gap between fuel and E100 prices declines to $P_{F1} - P_{E100}$ and so some E100 consumers switch to fuel consumption. As a result, the hydrous demand curve shifts in by the same amount (distance *e*) as the fuel demand curve shifts out (distance *b*); the distances are measured in gasoline energy-equivalent liters.

The reduced demand for hydrous ethanol results in a new demand curve D'_H, with a decline in the price of hydrous ethanol to P'_{E100}, thus partially offsetting the decline in hydrous ethanol consumption by distance *d* to yield a net reduction in hydrous ethanol consumption of distance $e - d$. If the fuel price stayed at P_{F1}, the resulting fuel consumption would correspond to C'_F. However, because the hydrous and anhydrous ethanol prices are linked on the supply side, the anhydrous ethanol falls, resulting in a further reduction in the fuel price, denoted by P_{F2}. Of the additional fuel consumption associated with this price decrease, 25 percent is anhydrous ethanol. In total, a reduction in the gasoline tax brought about an increase in fuel consumption of $a + b + c$, and, therefore, a higher need for anhydrous ethanol of $0.25 \times (a + b + c)$. On the other hand, the net decrease in the use of hydrous ethanol is $e - d$. Therefore, if $0.25 \times (a + b + c) < e - d$, the total use of ethanol declines, resulting in a decrease in both hydrous and anhydrous ethanol prices. A change in the gasoline price has qualitatively and quantitatively identical results as a change in the gasoline tax depicted in Figure 11.5.

With the explanation of the source of the ambiguity in mind, we next summarize a number of interesting qualitative and quantitative results for the effects of some other Brazilian policies on the ethanol and sugar prices.

The effect of a higher mandate on the market equilibrium is ambiguous, but de Gorter et al. (2013f) show that a higher mandate empirically results in a higher ethanol price. It is worth comparing this result with the effect of the US ethanol mandate in a situation when the gasoline price is not determined by market forces (i.e., is exogenous) as it has been the case in Brazil. In that specific case, a higher ethanol blend unambiguously results in a higher ethanol price. The reason for this difference is the fact that there are two mutually competing demands for ethanol, whereas the United States has only one demand for corn ethanol.

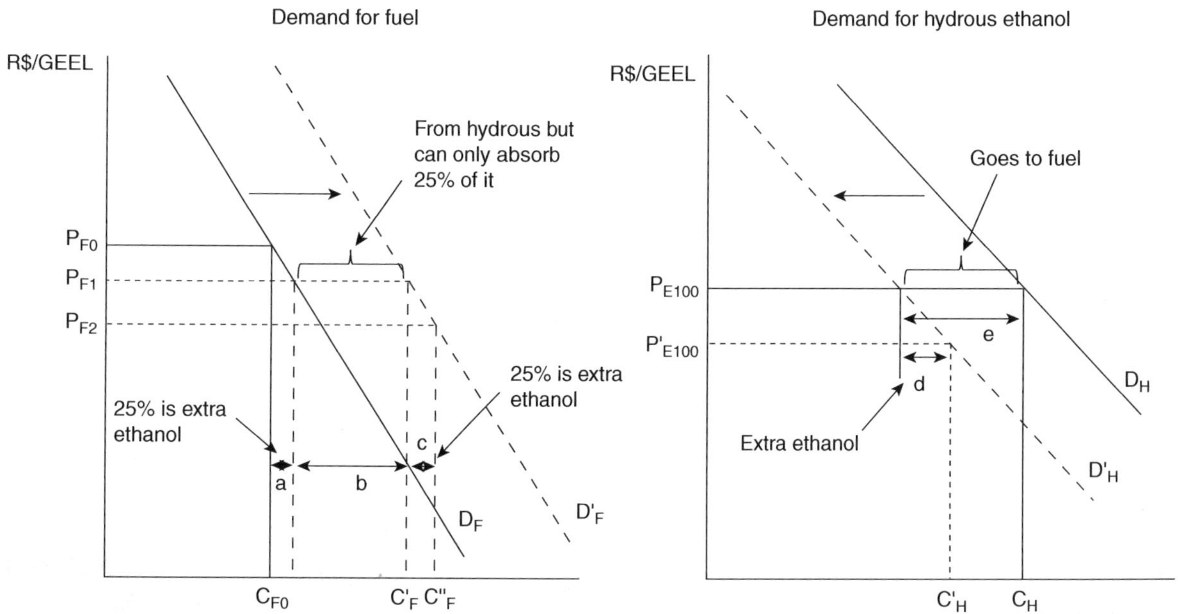

Figure 11.5 Effects of a decrease in Brazil's gasoline tax.

The tax on hydrous ethanol drives a wedge between the consumer and market price of hydrous ethanol. Because hydrous and anhydrous ethanol prices are linked, the latter decreases with a higher tax on hydrous ethanol (or equivalently with a reduction in a tax exemption). This makes the blending of the E25 fuel less costly which lowers the fuel price. Weaker competition for sugarcane depresses the feedstock's price and production. However, the reduction in the sugarcane use due to a lower need for ethanol more than offsets the reduction in sugarcane production, thus diverting more feedstock to sugar production. As the supply of sugar increases, its market price falls.

Interestingly, a higher tax on hydrous ethanol has an ambiguous effect on the consumer price of hydrous ethanol. The sign of the effect depends on the relative magnitudes of the inward shifts in the demand and supply curves for hydrous ethanol. The demand curve shifts in because of a change in the relative prices in favor of fuel while the supply curve contracts because a lower market price of hydrous ethanol makes this product less profitable to producers, who subsequently divert the feedstock (sugarcane) to sugar production.

In a previous section, we have identified several contributing factors to the significant rise in the ethanol and sugar prices in Brazil in the period 2010–2011 to 2011–2012. Brazilian biofuel policies were one of the factors. The question we would like to answer is to what extent these policies increased the ethanol and sugar prices.

The Monte Carlo simulations in de Gorter et al. (2013f) show that the total changes in biofuel policies that occurred between 2010–2011 and 2011–2012 had a negative and significant effect the observed market prices. In particular, in the absence of exogenous market shocks (e.g., a contraction on sugarcane supply because of bad weather), the anhydrous and hydrous market prices would decrease by approximately one third and the sugarcane prices by almost 40 percent.

This indicates that the adopted changes in policies are in line with the government's objective to combat inflation in the country. It is interesting to note, however, that, empirically, two policies that seemingly help the ethanol industry in Brazil do the opposite in reality: low gasoline taxes and high anhydrous tax exemptions lower ethanol prices. This suggests that the Brazilian government could be even more successful in reducing inflation if they did not lower the gasoline tax and did not increase the anhydrous tax exemptions.

How the Expiring US Tax Credit Pulled Brazilian Ethanol Prices Down

Curiously, changes in US ethanol policy may have been playing a role as well. Before it expired at the end of 2011, the US tax credit was paid on a majority of ethanol exports, thereby subsidizing US ethanol production (see the next chapter for an explanation). The expiration of the tax credit is a saving for US taxpayers, it reduces US ethanol profits and, thus, corn prices but is causing an increase in world sugar and ethanol prices, thereby hurting Brazilian fuel consumers. Meanwhile, the ever-expanding US ethanol mandate is diverting ethanol supplies that otherwise could go to world markets including Brazil, thereby exacerbating world sugar and ethanol market-price increases even more.

The importance of the EU trade policy, the US tax credit and the change in Brazil's market situation is illustrated in Figure 11.6. Brazilian and US ethanol prices followed each other and the US gasoline price until November 2011 when the European Union closed the tariff loophole[1] and the US tax credit expired at the end of December 2011. Both US and Brazilian ethanol prices collapsed, with the spread increasing (because the tax credit expired) and broke away from the US gasoline price. The connection between ethanol prices in Brazil and the United States links sugar prices to corn prices. By May 2012, Brazil no longer imported US ethanol and prices converged again. Developments after June 2012 are related to the US drought. However, Figure 11.6 illustrates how changes in biofuel and the interaction of biofuel policies significantly affect grain/oilseed prices. This has been the story all along since that fateful month of September 2006.

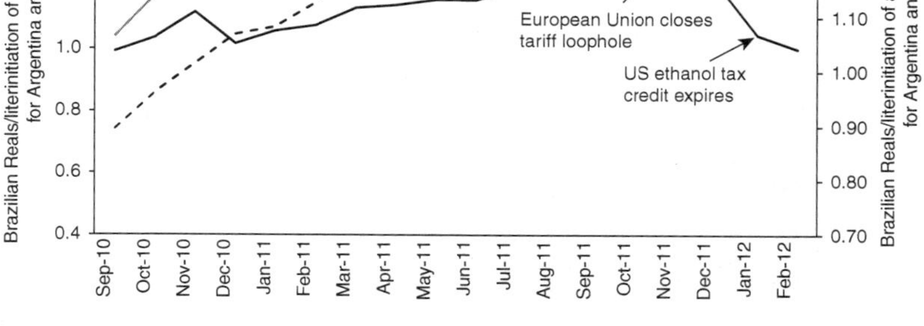

Figure 11.6 How the expired US tax credit dragged Brazil's ethanol prices down.

Note: *The value for April 2011 was truncated to equal 1.60 Brazilian Reals per liter (the actual value was 4 Brazilian Reals per liter).

Source: de Gorter et al. (2014).

Concluding Remarks

Brazil is a major player in world ethanol markets and, therefore, its policies can directly impact US ethanol and crop prices. Dramatic changes have occurred recently in Brazilian ethanol and sugar markets and policies. In order to combat inflation in late 2010, Brazil lowered the E25 mandate, increased gasoline prices, reduced gasoline taxes, and reduced the gasoline tax exemption for ethanol used in E100. However, E25 prices still increased by approximately 8 percent from 2010–2011 to 2011–2012, whereas ethanol, sugar, and sugarcane prices increased by more than 20 percent. This means supply/demand shocks dominated the markets.

In this chapter, we outlined a general economic model of the Brazilian sugar/ethanol nexus from the processing of sugarcane, where the world prices of sugar and ethanol are uniquely determined. We have stressed the importance of considering two demand curves for ethanol—one for fuel (a mixture of anhydrous ethanol and gasoline as E25) and the other for hydrous ethanol (E100)—as well as of incorporating an endogenous switching model for E100 consumers as they respond to changes in the relative price of E100 and E25.

Although the effects of policies taken together reduce prices, we show that, in theory, this needs not be the case. Unlike biofuel mandates and tax exemptions elsewhere, Brazil's fuel–ethanol–sugar markets and fuel policies are unique in that each policy, in theory, has an ambiguous impact on the market price of ethanol and hence on sugarcane and sugar prices.

The Brazilian market is complex with two mutually competing demands for ethanol; therefore, any initial change in ethanol price due to a policy change can be offset with shifts in demand for E100 versus fuel. Furthermore, the feedstock is sugarcane, which produces two competing products, sugar and ethanol, giving processors more flexibility.

Conventional wisdom is that higher levels of each of the mandate, tax exemptions for ethanol and gasoline prices, and a lower gasoline tax, all help the ethanol industry. However, for two of these policies, a low gasoline tax and a high gasoline tax exemption for ethanol used in E25, our empirical results show ethanol and sugar prices decline. Our theory and empirical framework gives a basis to predict market effects of individual policies and the effectiveness of Brazilian policymakers to achieve their goals.

Finally, our empirical estimates show that, for a given increase in ethanol prices, the 1.8-percent increase in the sugar price is significantly lower than the 5.8-percent increase for corn prices but more than the increase in soybean prices of 1 percent for the same increase in biodiesel prices. Therefore, sugarcane ethanol falls between corn-ethanol and biodiesel in terms of affecting food commodity prices.[2] However, there are three important points to keep in mind when assessing the relative food commodity effects of sugarcane- versus corn-ethanol and biodiesel. First, we assume all biodiesel is from soybean (rapeseed) oil yet a significant (and growing) amount of biodiesel is produced from recycled vegetable oils and animal fats. Therefore, we have overestimated the direct impact of biodiesel prices on food commodity prices. Second, the United States has corn equivalence factors ranging from 1.5 to 1.7 for biomass-based diesel.[3] The mandates are being filled with far less "wet" or actual physical gallons of biodiesel. Again, each gallon of biodiesel will have a lower impact on soybean prices than what we have estimated. Third, the law of one relative international price holds less for the sugar price than for grains and oilseed prices because the degree of competition for land and substitution in demand is less for sugarcane and sugar, respectively. Therefore, the impact of sugarcane-ethanol on food commodity prices is also expected to be less than what we have estimated.

Having said all of that, what happens if one fixes the amount of biofuel (not price per gallon) in comparing the price effects across feedstocks? One can produce a lot of food using an acre of corn that produces seven times the biofuel on an acre of soybeans[4] and grow corn or wheat for human consumption on the seven (or 4.7) soybean acres instead. Therefore, much more research has to be done in ascertaining the relative efficiency of different biofuel feedstocks on food commodity prices.

CHAPTER 12

THE INTERACTION EFFECTS AMONG BIOFUELS, POLICIES, AND COUNTRIES

The models developed in this book span different feedstocks, biofuels, and countries and, therefore, we have already come across several important interactions. For example, we closed the previous chapter on how the expiring US tax credit impacted Brazilian ethanol prices. In the chapter before that, we showed how the complex structure of the US mandate affected domestic and international ethanol and biodiesel prices and trade. Throughout the book so far, we showed how market shocks or policy changes can affect regime changes (e.g., the mandate versus tax credit binding). These interactions can occur across biofuel policies, within a country and across countries and biofuels.

We also showed how the US "splash and dash" policy with the biodiesel tax credit affected international prices or that Argentina with differential export taxes between soybeans and biodiesel affects world biodiesel prices. We detailed how the law of one relative international price has the relative grains and oilseed prices do not change over time because of competition for the same land and substitution in demand. We even developed a formula in Chapter 2 [equation (2.1Y)] that showed how extracting corn oil from the dried distillers grains with solubles (DDGS) and selling it for biodiesel production increases corn prices significantly, further integrating corn and soybean markets. It also appears that biofuel policies have further integrated corn and wheat markets and prices as the DDGS and wheat prices are positively related (see Figure 12.1). This is because a good deal of wheat is sold as animal feed (we noted earlier that feed wheat consumption increased from 2007 to 2010 by more than twice the production shortfall of wheat due to bad weather in crop year 2007–2008 that so many economists allege caused the wheat price run-up in 2008) and wheat contains more protein, like DDGS, compared to corn. Therefore, this positive relationship is expected.

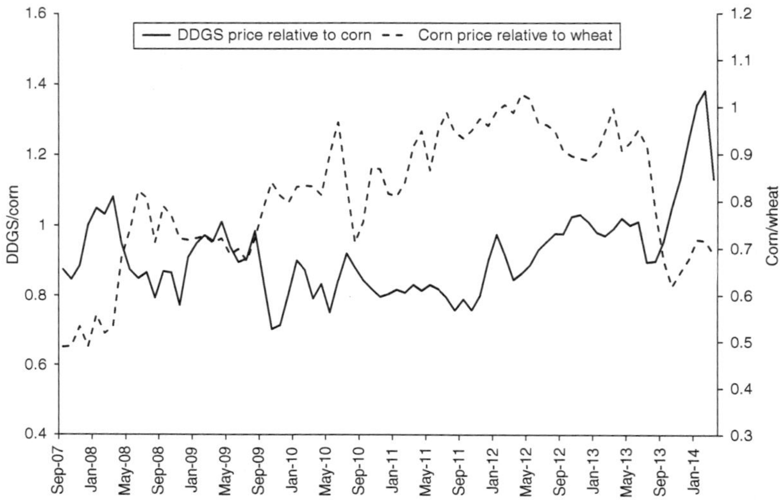

Figure 12.1 How wheat and DDGS prices are positively related.
Source: AMRC (2014), World Bank (2014).

Thus, the reader is very familiar with interaction effects discussed so far; therefore, it is somewhat anticlimactic to introduce an entire chapter on the topic at this point in time. But it is necessary, if only to outline some guiding principles and implications of our general theory of biofuel policies. Therefore, the purpose of this chapter is to highlight the basic economics of interaction effects between biofuel policies among themselves (mandates, tax credits, and ethanol-production subsidies), farm subsidies, and technology including R&D policies. Biofuel policies have also caused new political interactions: the new politics of the US farm bill has been affected by biofuel policies. Finally, there are irreversibilities created by biofuel policies; the effect of implementing a biofuel policy is not the same as taking the policy away once it is in place. We now turn to describing all of these interaction effects. It should be noted that the effects of a tax credit with a binding mandate, for example, are different in a closed economy than if, for example, the United States is exporting ethanol. Therefore, we need to distinguish open economy from closed economy. We begin with the latter case.

Closed Economy

In previous chapters, we have discussed the economics of the two most frequently used biofuel policies worldwide—the biofuel blend mandate and the blender's tax credit in the United States (or the excise tax exemption in the rest of the world). We have focused on the effects of these instruments one at a time. However, in reality, blend mandates have historically been complemented with tax credits (tax exemptions); biofuel- and crop-production subsidies have been in place; moreover, a country's biofuel policies have implications for the effects of policies in other countries. Therefore, it is crucial to understand not only the interactions between biofuel policies but also between biofuel policies and farm programs, as well as the international policy effects. To be able to trace these interaction effects, we only look at a combination of two policies at a time.

A Binding Blend Mandate Combined with Other Consumption and Production Subsidies

Perhaps the most discussed policy combination in the United States has been the blend mandate and the tax credit. The provision of the blender's tax credit has been justified on the grounds of incentivizing fuel blenders to meet the biofuel mandate. It has also been argued that the tax credit and the mandate price premium are additive, meaning that the corn prices increase by the full amount of the tax credit (in $ per bushel) as compared to the situation under the mandate alone. However, de Gorter and Just (2009b) showed that this is not the case and that the effect of an ethanol tax credit added to a binding mandate is (mostly) to subsidize the gasoline consumption (but if the tax credit is binding, the blend mandate is redundant). This is an excellent example of often contradictory interaction effects of biofuel policies (from the point of view of policies' objectives).

To understand this interaction effect, recall that a blender's tax credit is an ethanol consumption subsidy. Therefore, its primary effect is to reduce the fuel price, thus increasing fuel consumption (ethanol plus gasoline). The increase the quantity consumed is small, however, because the fuel demand is price inelastic; moreover, the fuel price does not decrease by the full amount of the tax credit but only by a fraction which corresponds to the share of the biofuel. Because fuel consumption increases, so must also ethanol production due to the existing mandate which drives up the ethanol price. However, the increase in the ethanol price is small because of the inelasticity of fuel demand and because the ethanol supply curve is more elastic than assumed (due to the recycling effect). This means the ethanol price increases just by a fraction of the actual tax credit, making the mandate price premium and the tax credit not additive.

Therefore, what is then the incidence of the tax credit? Fuel blenders can take advantage of the tax credit only when they blend more ethanol. Because the quantity of ethanol is firmly linked to the quantity of fuel, the only way blenders can benefit from the tax credit is to voluntarily lower the fuel price, thus obtaining a bigger market share and being able to sell more fuel. The competing blenders will keep lowering the fuel price and bidding up the ethanol price until they make zero marginal profits. As a result, the tax credit with a binding mandate leads to higher gasoline consumption, contradicting all three pillars of policy goals biofuels seek to fulfill: reduce energy dependence, improve the environment, and improve farm incomes.

From a distributional point of view, the taxpayer's money collected for the tax credit becomes a transfer to fuel consumers and gasoline producers. Gasoline producers gain from higher production. Being directly linked to the ethanol price, the corn price also increases when a tax credit is combined with a binding mandate, although this increase is very small.

However, the tax credit is not the only subsidy the US biofuel industry has historically benefited from. There are currently also ethanol production subsidies, as well as feedstock production subsidies (e.g., for corn and soybeans) that lower the cost of the biofuel feedstock.

The ethanol-production subsidy lowers the marginal cost of ethanol production and, thereby, the ethanol market price. This makes the fuel blend less costly for consumers and the fuel consumption increases. In this respect, ethanol production subsidy has the same effect as an ethanol blender's tax credit (a consumption subsidy). The ethanol production subsidy drives a wedge between the ethanol producer price (in other words, the market price plus the production subsidy; this price is linked to the corn price) and the market price of ethanol (faced by fuel blenders). The increase in the corn price due to ethanol production subsidy is likely to be small because of inelastic demand for fuel and a relatively elastic ethanol supply curve.

The immediate effect of a corn production subsidy is to expand corn production, thus, by extension, also the availability of corn for ethanol production. This lowers the marginal cost of ethanol production and consequently also the cost for fuel blenders. Thus, akin to the tax credit and the ethanol production subsidy, the corn production subsidy implicitly subsidizes fuel consumption when combined with a binding blend mandate. Because, in equilibrium, quantities of fuel and ethanol are linked through a blend mandate, production of ethanol increases but the market price is actually lower because of lower marginal cost of ethanol production. Owing to the link between ethanol and corn prices, consumers of corn for non-ethanol and ethanol use enjoy lower market prices; however, corn producers receive the market price plus the subsidy, thus earning a higher price as compared to the situation without subsidy.

As summarized in Table 12.1, all three subsidies result in perverse effects when combined with a binding mandate: they reduce the consumer fuel price which implies higher fuel and, therefore, also higher gasoline

Table 12.1 Comparative statics results for a binding blend mandate combined with other subsidies

		Effect on price of:			
		Fuel	*Gasoline*	*Ethanol*	*Corn*
Change in	Tax credit	–	+	+	+
	Corn-production subsidy	–	+	–	–
	Ethanol-production subsidy	–	+	–	+

Source: Drabik (2011).

consumption. This contradicts the objectives of reducing energy dependence, improving the environment, and even helping farmers.

A Binding Tax Credit Combined with Corn and Ethanol Production Subsidies

In periods when the ethanol mandate was either not in place or was dormant (in other words, not effective), the blender's tax credit determined the ethanol market price (and the ethanol and hence corn prices were lowest when locked onto the crude oil prices). It is, therefore, important to investigate the interaction effects of ethanol and corn production subsidies under the tax credit.

An ethanol production subsidy reduces the marginal cost of ethanol production, which translates into a lower market price of ethanol (paid by fuel blenders), but ethanol producers receive a price equal to the market price plus the full amount of the subsidy. It should be noted that the reduction in the ethanol price is smaller than it would be under the binding mandate because, in the present case, the ethanol price is locked on to the crude oil price, which does not change much (if at all) with ethanol production. In a competitive ethanol-producing environment, the subsidy is a transfer to corn producers who expand their production. On the other hand, consumers of corn for feed/food are worse-off because of an increase in the corn market price.

The ethanol-production subsidy under the binding tax credit reduces the consumer fuel price to the extent to which the additional ethanol production due to the subsidy can depress gasoline prices.

Now consider a corn-production subsidy. Akin to the situation with the mandate, the subsidy lowers the marginal cost of corn and ethanol production. To simplify the matters, let us assume that the crude oil price and, therefore, the gasoline price is fixed; the same then holds for the ethanol and corn prices [due to the unique link given by equation (2.1) in Chapter 2] if the tax credit is binding. The effect of the corn production subsidy in this case is to expand corn and ethanol production. But, why should ethanol producers produce more ethanol if they receive the same market price?

Because the ethanol price does not change while the corn-production subsidy has lowered the marginal cost of ethanol production, ethanol producers initially earn windfall profits. This implies that new producers will enter the market and produce more ethanol, thus consuming more corn; alternatively, the incumbent producers may expand their production. Competition in the ethanol production industry ensures that the producers bid up the price of corn until the marginal firm earns zero profits again.

Because the corn market price does not change (much) with the corn-production subsidy under the binding tax credit, so does not consumption of corn for feed/food use. This situation motivates the notion of the recycling effect (explained in Chapter 2) because it is probably the only explanation how the corn and ethanol markets can be in equilibrium under the conditions described earlier. The additional quantity of corn produced due to the corn production subsidy shifts to ethanol production, complemented by the yellow corn obtained from the change in the composition of non-ethanol corn use caused by the additional quantity of the DDGS, which was induced by the corn-production subsidy.

As before, by lowering the marginal cost of ethanol production, the corn-production subsidy reduces the consumer fuel price (but only to the extent to which the expanded ethanol production affects the crude oil price. Corn producers receive the market price of corn plus the corn-production subsidy. Note also that because the corn production subsidy expands ethanol production, more co-product is returned to the corn market which crowds out yellow corn from feed/food consumption.

In summary, the tax credit by itself reduces gasoline and fuel prices, whereas it increases the ethanol and corn prices. This happens because the tax credit induces higher ethanol production and, therefore, also higher corn production. On the other hand, ethanol crowds out gasoline whose production declines; therefore, the decrease in the gasoline price. Under a binding tax credit, the fuel price equals the sum of the gasoline price and the fuel tax because this is what consumers are willing to pay. The ethanol production subsidy, by reducing the ethanol market price, lowers the marginal cost to fuel blenders, while expanding ethanol production because the producers receive the ethanol market price plus the subsidy. The corn price increases because it is linked to the price received by ethanol producers. A corn-production subsidy reduces all market prices analyzed. This is because it lowers the marginal cost of corn production, thereby expanding ethanol production as the former becomes less costly. Table 12.2 presents a succinct summary of these results.

Table 12.2 Comparative statics results for a binding tax credit

		Effect on prices of:			
		Gasoline	*Ethanol*	*Fuel*	*Corn*
Change in	Tax credit	–	+	–	+
	Corn subsidy	–	–	–	–
	Ethanol subsidy	–	–	–	+

Source: Drabik (2011).

Given the two binding biofuel policies considered earlier—the mandate and the tax credit—and the unique links between ethanol (and therefore corn) and gasoline prices corresponding to these policies, it is important to pose a question on how a binding biofuel policy affects the impact of alternative subsidies on corn prices. We illustrate these effects using the results of the model in Drabik (2011) for the year 2010.

If a tax credit is binding, corn-production subsidies have minimal (but negative) impacts on corn market prices as any impact has to go through oil prices; however, under a mandate, corn prices fall by 25 percent more as compared to a case where a corn-production subsidy were provided in a situation with no biofuel production. In contrast, the (observed) ethanol production subsidy alone is not able to induce ethanol production because of the high levels of redundancy in the ethanol price premium due to ethanol policy (what we earlier called "water"—see Chapter 2 for a detailed explanation), and only negligibly increases corn prices when in combination with the mandate (which determines the ethanol, and hence corn price). However, the ethanol production subsidy has a full effect in increasing corn prices (and has a negligible impact on ethanol market prices) when in combination with a binding tax credit because the subsidy drives a wedge between ethanol producers' and market prices (paid by blenders).

The previous example illustrates the need to correctly calibrate the model to the binding biofuel policy (combined with other policies in place) for simulating the combined effects of biofuel and agricultural policies. As referred to in Chapter 2, a model calibrated to differing binding biofuel policies can easily produce different and even opposite policy recommendations (e.g., on the optimal level of the blender's tax credit or mandate).

International Interaction Effects

So far, we have focused our discussion on the interaction effects of biofuel policies in one country only. However, the global biofuel markets are interlinked and countries use different policies to promote consumption and production of biofuels. In this section, we use the example of biodiesel policies in the United States and the European Union to show their historical international interaction effects that materialize in so-called "splash and dash" trade dispute.

In a closed economy, a blender's tax credit has the same effect as an excise fuel tax exemption—both constitute a biofuel consumption subsidy that raises the market price of the biofuel and, thereby, establishes an advantage over gasoline/diesel production (de Gorter and Just 2008b, 2009a).

The effects of these two policies differ, however, in an open economy framework, depending on which policy determines the world price of a biofuel (Kliauga et al. 2011; Rajcaniova et al. 2013) and which policies are in place in other countries. Either policy (a tax exemption or a tax credit) in the price-determining country is an actionable subsidy by itself, from the point of view of oil producers (but not for biofuel producers, as biodiesel producers gain worldwide). It is because either the tax exemption or the tax credit, depending on which determines the world market price, increases the market price of biodiesel; therefore, its production is above a level that would have otherwise existed. Therefore, oil production may be crowded out due to these policies.

To ease the understanding, consider only two countries (e.g., the European Union and the United States), each producing biodiesel. The question now becomes whether a tax credit (tax exemption) in the other country is also necessarily an actionable subsidy from the point of view of oil producers. Not always. The reason is that once the world market price of biodiesel is established by one country (A), a tax credit or a tax exemption in the other country (B) cannot affect it, but acts as a production subsidy in the case of a tax credit and a fuel consumption subsidy with a tax exemption. Therefore, if country B has a tax credit, then it is an actionable subsidy, but a tax exemption is not (oil producers can even benefit from it because fuel and, thus, diesel consumption increases).

Furthermore, if country B has a tax credit (as was the case with the United States and the "splash and dash" program), it expands biodiesel exports as a result of increased biodiesel production and imports from third countries that are "splashed" and then "dashed" to country A (the European Union).[1] If the United States determined the world biodiesel price (country A in this case) through a tax exemption, then the EU tax exemption would subsidize fuel consumption and would have little (or no) effect on US biodiesel prices. On the other hand, if the European Union had a tax credit, it would subsidize biodiesel production.

One implication is that the entire "splash and dash" controversy as described in Chapter 9 earlier (see de Gorter et al. 2011 for details) would not have existed had the European Union switched their tax exemption to a blender's tax credit, or had the United States switched their tax credit to a tax exemption. Thus, no country could "double dip."

Technology

One of the alleged goals of biofuel policy is to induce technological change (Rajagopal et al. 2009; Fischer and Newell 2008). In this section, we show how biofuel policies mediate interesting interaction effects between technology and feedstock (e.g., corn) prices. We show that the effects of a technological change on corn prices depend on (1) the binding biofuel policy: the blender's tax credit or the mandate; and (2) the source of the technological change.

Our simple analytical model for corn ethanol presented in Chapter 2 allows us to model two sources of a technological change. First, a change in the number of gallons of ethanol from a bushel of corn, β, can for example, represent advancements in the crop breeding. Second, developments in new technologies that reduce the ethanol processing cost, such as the hydraulic fracturing that reduces natural gas prices (the highest non-corn cost of ethanol processing) are captured in the parameter c_0 in equation (2.1).

Assume the tax credit determines the ethanol market price. In this case, our model predicts[2] that, regardless of the source of the technological change (in other words, either more ethanol is extracted per bushel of corn or ethanol production becomes less costly), the corn price is expected to increase. This happens because the new efficiencies in ethanol production allow ethanol producers to bid up the corn price (for the same ethanol price). Gardner (2007) arrives at the same qualitative conclusion, albeit using a different model set-up.

Now, let us assume the mandate binds. In this case, the model produces mixed results. A higher extraction rate of ethanol from corn (in other words, an increase in the parameter β) can either increase or reduce corn prices. However, the reduction in the processing cost always increases the corn prices. This implies that when the mandate binds, the source of the technological change matters even more.

However, the technological change in agricultural production (Sexton and Zilberman 2012), such as higher corn yields per hectare, could reduce crop prices under either policy, albeit to a different extent. Under the mandate, the expansion in corn production would lower the marginal cost of ethanol production, thus lowering its price as well as the consumer fuel price. Being linked to ethanol prices, corn prices would decline as well. However, once the mandate premium dissipates, crop prices are tied to crude oil prices through the tax credit such that improved agricultural productivity can only reduce crop prices by increasing biofuel production sufficiently to reduce crude oil prices. In that scenario, the technological change in crop production would lead to minimal changes in the ethanol and, thus, corn prices (because of the specific link between them) because the additional corn produced would be processed into ethanol.

The overall implication of the above findings is that improved technology (related to ethanol processing) of either type considered will not necessarily solve the food–fuel price trade-off.

An Example of Interaction Effects Between Commodities

New biofuel production technologies also create indirect interaction effects between feedstocks used to produce biofuels. As an example, consider the extraction of corn oil from DDGS (a co-product of ethanol production) that is later used for biodiesel production. Because this corn oil increases corn prices, on average, by 18 cents per bushel [as per equation (2.1′)], the developments in the corn oil (in other words, biodiesel) market directly affects corn market prices. Because the corn oil price is mainly affected by the price of diesel, for whose production the corn oil is used, we have a situation in which a commodity (in other words, corn) price can be determined by a seemingly unrelated product—the biodiesel price. It also is an example of how biofuel policies have made corn and soybean markets and prices more integrated; not just through the law of one relative international price with competition for land and substitution in demand.

Political Interactions

Biofuel policies and the resulting high and volatile grain and oilseed prices have created a new politics of the US farm bill. Moreover, the central role energy and environmental policy play in determining grain prices has also changed the political–economic landscape that directly affects grain prices, so that now US farm legislation wields less impact than it did previously. The major stakeholder groups who have been engaged in the political process and the specialized interest-group landscape have changed since 2006. There is no longer an "iron triangle" that influences commodity prices; instead, we now have the "iron maze" of environmental, energy, and agricultural legislation, interest groups, and public agencies (Rausser and de Gorter 2012).

Because ethanol policies support feed-grain markets through higher prices, livestock, dairy, and poultry producers are beginning to form organized opposition interest-groups, as are those food processors that no longer enjoy the low market prices that traditional agricultural policies ensured. New interest groups include coalitions of livestock organizations such as various meat, livestock, poultry, and dairy producer associations. Because of both higher feed costs and volatility in milk prices, new proposals for "Margin Protection" type programs with voluntary supply controls are the central features of dairy policy reforms. Although the federal budget and deficit woes are to shape the ultimate program design, "cuts" are related to the Congressional Budget Office (CBO) baseline and "mandatory" spending, not current or recent actual expenditures.

The higher commodity prices due to biofuel policies have resulted in a new politics of the farm bill where direct payments are embarrassing and, therefore, were eliminated, along with the counter-cyclical program. Most of the so-called reforms are largely cynical, because in this new era of high prices and record farm incomes, direct payments (where most of the "cuts" fall) were politically unviable and likely to end anyway; and counter-cyclical and loan-deficiency payments (which are paid out when prices are low) are unlikely to be triggered. Therefore, the new farm bill introduced the "price loss coverage" (PLC) program with a system of much higher target prices, generating production distortions with implications for crop production, commodity prices, federal outlays, the environment, and US international trade commitments.

Meanwhile, in an era of high prices, subsidies associated with the crop insurance programs have become very high, especially if one includes underwriting and administration costs. With high prices, crop insurance subsidies in recent years have, therefore, become the biggest spending category (approximately $7 billion for total US crops). The political desire to transfer income to farmers in high/volatile price era remains, but farm policy has to change. Therefore, revenue insurance (or "shallow loss" protection with crop insurance programs) is the new mantra in Washington. This is one way of ensuring farmers get subsidies with high and volatile crop prices. The new Farm Bill introduced "revenue insurance" programs that cover "shallow losses" of crop insurance designed to complement crop insurance. The most important of these is the Supplemental Coverage Option (SCO) shallow loss program. Evidently, Farm Bill advocates believe it is more advantageous to their interests to redesign such policies as "revenue-insurance" programs that will kick in whenever prices vary (and the loan rate serves as a minimum price guarantee).

The New Economics of Impacts of Subsidies and International Trade Distortions

If a biofuel consumption subsidy (the blender's tax credit in the United States or excise tax exemption in the rest of the world) is binding, corn-production subsidies have minimal (but negative) impacts on corn market prices as any impact has to go through crude oil prices; however, under a mandate, corn prices fall by 25 percent more as compared to a corn production subsidy in a situation with no biofuel production (Drabik 2011). Therefore, the traditional effects of farm subsidies are now quite different and have implications for World Trade Organization (WTO) trade disputes.

An ethanol-production subsidy, on the other hand, reduces the ethanol market price, lowers the marginal cost to fuel blenders, while expanding ethanol production because the producers receive the ethanol market price plus the subsidy. The corn price increases because it is linked to the price received by ethanol producers. But a corn-production subsidy reduces all market prices analyzed. This is because it lowers the marginal cost of corn production, thereby expanding ethanol production as the former becomes less costly. Therefore, the market distortions of biofuel policies are to increase market prices of grains/oilseeds (something the WTO does not concern itself with) whereas traditional farm and trade policy impacts on world prices are now different (compared to the previous era of no biofuels), with world price distortions depend on the biofuel policy regime (world prices distorted more (less) with a mandate (tax exemption) determining the biofuel price compared to the era of no biofuels).

Irreversible Investment Interactions

Perhaps the most important implication of biofuel policies, apart from creating the crop–biofuel price links with high multipliers, is the irreversible nature of biofuel policies: the impact of removing all of these policies now is very different than not having the policies in the first place.

This is true for all of the investments in ethanol production, ethanol blending, and ethanol distribution where billions of dollars of fixed costs have been sunk. Investments were aided by taxpayers in the form of consumptions subsidies for biofuels and production subsidies for biofuels and the feedstocks, transfers from fuel consumers, and from investors where the value of publically and privately held shares plummeted from their 2008 peaks. It has been estimated that 50 percent of ethanol refineries went bankrupt in 2009 and were bought for dimes on the dollar. Therefore, all of these transfers represent sunk costs so removing the policies today will not reduce biofuel production as much as these transfers initially promoted biofuels production.

A classic example of this phenomenon is the irreversible investments made in refining crude oil because refiners saw the writing on the wall that a 10 percent ethanol blend was inevitable because of the renewable fuel standard (RFS). Gasoline consumption was forecast to decline and, therefore, to maximize profits, refiners made irreversible investments to use ethanol as an octane enhancer. This is adequately detailed in Irwin and Good (2012), Babcock (2012b), and Tyner et al. (2012). Amazingly, eliminating the RFS in the drought-stricken year 2012 allegedly would not have any impact unless ethanol prices would have exceeded $3.40 per gallon, which translates into a $10 per bushel corn price (see Drabik and de Gorter 2012). Irreversible investment narrows the short-run policy options and should corn prices have gone toward $10 per bushel, clearly the policy advice should have been to permanently waive the mandate *plus* a reverse mandate not allowing corn-ethanol as an octane enhancer to undo past damage (Drabik and de Gorter 2012). In a way, one cannot simply eliminate food–biofuel competition because it is irreversible *ex post*. This has important implications for policy advice in both the short and long run.

Concluding Remarks

This chapter has focused on interaction effects among biofuel and farm policies both domestically and internationally. The main conclusion for a mix of biofuel policies is that, if a binding mandate is combined with any combination of tax credits (excise tax exemptions), biofuel and feedstock (e.g., corn or soybean) production subsidies, the resulting effect of the non-mandate policies is to subsidize fuel consumption, thus contradicting the main policy goals biofuels seek to fulfill: reduce energy dependence and improve the environment and improve farm incomes. We have also shown that a corn-production subsidy has a considerably bigger effect (empirically we estimate 25 percent bigger) in reducing corn prices with a binding ethanol mandate than if there were no biofuel policies at all. In addition, we have shown that, although both the corn and ethanol-production subsidy under a binding ethanol mandate result in higher corn prices received by corn producers, the channels through which this happens differ; this brings us into a completely new environment of analyzing the impact of biofuel and agricultural policies.

The interaction effects among biofuel and agricultural policies in a closed economy become even more complex in an open economy framework. To determine the market effects of individual policies, it is necessary to know the policy of which country determines the biofuel price. Finally, we illustrate that biofuels have triggered interaction effect not only among policies but also between commodities and have affected politics as well as investment decisions of fuel blenders.

Because our discussion in this chapter focuses only on first-generation biofuels, it is imperative that future research also analyze the economics of biofuel policies of second-generation biofuels, particularly the interaction effects of the subsidies for second generation biofuels with the first generation tax credits (tax exemptions) and farm subsidies on agricultural crops, such as corn. The priorities for second-generation biofuels may well be influenced by the impact of first-generation policies on food commodity prices. There are important policy issues surrounding second-generation biofuels (produced from crop biomass residues and from dedicated cellulosic crops) in terms of food commodity prices. The food versus fuel debate could be diffused if, in the future, there is not only a switch to second-generation biofuels produced from biomass not produced for food, but from residual biomass [from current crop production (e.g., corn stover) rather than from dedicated cellulosic crops (e.g., miscanthus) that could compete for land that now produces food]. The economics of each will differ with residual crop biomass have the biggest potential to relieve biofuels impact on food commodity prices. If crop residue could be used for biofuels (cellulosic or second-generation biofuel), then the price-increasing effects of biofuels policy would be moderated (if not fully offset) and less (if any) land will have to be converted from grasslands or forests. Likely, more land will be allocated to corn production as the value of the corn crop goes up initially because corn stover now has value. With more land devoted to corn production means the price of grain corn declines for a given initial level of ethanol (provided there are mandate premiums; otherwise corn prices are tied to crude oil prices and changes in ethanol production will have little impact on crude oil prices). However, presumably second-generation ethanol will displace some grain corn-ethanol; therefore, corn prices may decline even further. However, even if land is pulled out of other crops, there will be a moderating effect as all crops are substitutes in demand as well. The net effect on grain corn prices will depend on the profitability of corn stover ethanol, the amount per hectare produced, the initial proportion of corn going into ethanol, and the demand elasticity facing ethanol. The latter critically depends on whether the mandate is binding or corn prices are tied directly to crude oil prices through ethanol and gasoline prices. However, clearly, biomass from dedicated crops would not lower grain corn prices as much as that from crop residue.

CHAPTER 13

THE IMPACT OF BIOFUEL POLICIES IN THE FUTURE: SOME CONCLUDING REMARKS

This Is Only the End of the Beginning

We have analyzed the great food commodity price boom since late 2006 using a strict meat-and-potatoes approach. We do not discuss tulip manias or Malthus' revenge. We do not wax poetic about how Henry Ford had ethanol as his fuel of choice for the Model-T (this time is different) or that prohibition was a conspiracy to monopolize the gasoline market. We did not use the fanciest statistical techniques known to mankind nor did we use Big Data—we were not flashy and, therefore, minimized the use of algos and routers. We did not run numerical simulations with potentially unbounded price expectations to generate bubble like price behavior. Nor did we indulge in instant economics or take shortcuts. We take the straight and narrow path to the core of the issue: the economics of blend mandates and the new price linkages between crops and biofuels and between biofuels and energy (gasoline/diesel and crude oil) prices where sectoral supply/demand shocks have a very different impact on food commodity prices than before. In doing so, we just stuck to the basics—an application of microeconomics to biofuels production and consumption, the price and quantity links backwards to the feedstocks and forward to energy and crude oil markets, with due regard to co-products, joint products and by-products along the way. We simply assume profit-maximizing agents under competition. The results are powerful and insightful, although the models are a simplification of reality yet necessary to understand what caused deviations from our model predictions.

We have taken a very strict biofuel policy centric approach in this book: we develop the basic economics of biofuel policies for the three basic biofuels: corn-ethanol, soybean- and rapeseed oil-biodiesel, and Brazilian sugarcane-ethanol. We have analyzed biofuel policies in many dimensions, including various combinations of interactions across policies, biofuels, feedstocks, and countries. This book strictly adheres to a positive analysis of price determination. In no way have we undertaken normative policy analysis to answer such questions as whether or not higher prices or more price volatility is good for the world's poor, who spend a majority of their income on the basic foodstuffs we analyzed in this book. However, we do argue that one has to understand the exact underlying forces that caused the food commodity price boom before appropriate policy analysis can be undertaken. Furthermore, we believe we have advanced the correct framework of analysis. Our analysis in Chapter 6 on the effects of developing country policy responses in 2008 is only one example of the potential importance of our framework for policy analysis.

Although we do not undertake either prescriptive or predictive analysis, we do, however, speculate—see the next section. But so far, we just tell it the way it was and is at the time of writing; eight years of turmoil in the grains/oilseeds markets that are still not well understood by the profession largely because the profession has been more interested in using traditional methods and has resisted using the proper models and getting it right. The great commodity boom that reached one peak in 2008 was viewed by some as a silent tsunami (The Economist 2008a). But it was not silent; it just looked that way because economists, including *The Economist*, were looking at the wrong things; the things they always looked at in the past. However, this time was different.

There is or was no "perfect storm"; it was just corn (and other grains and oilseed) prices that were directly and indirectly affected by biofuel policies that, for the first time ever, created a price link between crops and biofuels with very high multipliers (especially for corn and wheat). And how long does a storm last? And how long does perfection last? There was only one unexpected shock to the markets from late 2006 to now: biofuel policies. Markets made a "mistake" not predicting that high crude oil prices along with the 1978 tax credit and various environmental regulations (along with the guarantee underwritten by the mandates) would cause the crop–biofuel price linkages. There was not a "perfect storm" of mistakes; the markets knew full well that China's gross domestic product was growing over 9 percent per year since 1978, that climate change had no impact on crop yields (in fact, the variance in crop yields has dropped), that crop yields in Africa accelerated in the past decade, that some flooding here and some drought there are perfectly normal events, and so on and so forth. One cannot presume that markets are stupid or prone to make mistake after mistake. However, markets did fail to anticipate the effects of biofuel policies, just as markets failed to predict the United States going off the gold standard and the Soviet Union suddenly entering the world wheat markets in the early 1970s. Shocks of this nature apparently happen every 30–40 years or so.

But this shock is different; it has lasted a lot longer than any other price boom period. This shock has long run implications, and, yet, the future is uncertain (see the following section where we reverse the title of this section!). Agricultural economists have always argued the best cure for a disease is the disease itself. That is to say, the best cure for high prices is high prices, which then lead to greater quantity supplied and lowers quantity demanded and a moderation of prices. And presumably vice versa; the best cure for real declining prices as shown in Figure 1.3 is low prices. However, was this the cause of the price boom that first emerged in September 2006 or was it biofuel policies? Some argue that grain and oilseed markets were due for a correction. But we argue that there was a very specific cause for the price run-up. And we argue that high prices have so far not been the cure for high prices. This is because the crop–biofuel price links have yet to be broken. Crude oil prices would need to decline significantly to break this link. Either traditional price declines or the breaking of the oil price link are possible. But so far, finding normal grain and oilseed prices again has not been as easy as it might seem given the "perfect storm" paradigm. This is especially true if crude oil prices climb and the blend wall worldwide is overcome. Unlike previous agricultural price booms and busts, we have so far missed the bust phase, indicating that this time may indeed be different. There may be a reason for that. It is not Midwest flooding, a freeze in Mexico or dryness in China causing price booms but rather the economics of biofuel policies and their effects on the grains/oilseed markets; in other words, everything that we have presented in this book.

By the end of 2008, some economists argued biofuels was just a fad. A fad is defined as "an intense and widely shared enthusiasm for something, esp. one that is short-lived; a craze."[1] But this book raises the question: has the long-lived framework of analysis used by most outlived its usefulness in this new era? Since September 2006, we have entered a new era of biofuels where there is a price link between food grain/oilseed and biofuel (ethanol and biodiesel) prices and another link from biofuels to energy (gasoline and diesel) prices. We argue that traditional analysis that completely ignores our framework (Chavas et al. 2014; Wright 2014; and Tomek and Kaiser 2014, being recent examples)[2] is not an option, but the reverse is also not an option; we cannot ignore the traditional models upon which all of agricultural economics has been based. We are not repealing the laws of supply and demand, macroeconomics, the theory of speculation, and the competitive theory of storage. Midwest flooding, dryness in China, or any other "perfect storm" explanation economists focus on must still have their impact. Therefore, these two approaches need to be integrated: the new with the old. This is the first order of action for the future. However, the economic framework of biofuel policy, as we have developed in this book, has to be the pivot point; the point of departure for any useful analysis to be done.

And yet, the analysis in this book is only the end of the beginning. In no way is this book the final word. Much more research has to be done. The rate of adoption of our approach by the profession has been glacial[3] and we have no idea why; it is beyond the scope of this book to answer that question. In a discussion of the book *Homo Academicus* written by the late Pierre Bourdieux, it was noted that[4]

> [A]cademics too often participate in a kind of mutual reassurance scheme: cite me and I'll cite you; praise me and I'll praise you; be clever and facile but do not be too demanding because most of your colleagues want new understanding much less than they seek comforting reassurance that they already know everything worth knowing.

This book, however, could have been easily entitled *Homo Economicus*, depending on how one specifies the objective functions of universities and the incentive structures of the economics profession. However, the economics profession responded with a myriad of literatures (we classified and reviewed four general categories of these other literatures in Chapters 5–8) in trying to explain the impact of biofuel policies. Each of these literatures, contradicting each other in methodology and in outcomes, reflects the chaotic quagmire of economic analysis of agricultural commodity markets [as the report of the High Level Panel of Experts on Food Security and Nutrition (HLPE) (2013) found out!]. Without developing new tools and relying instead on the unsuitable conceptual frameworks based in a prior reality to explain the price boom and volatility could severely impact the credibility of our profession.

Or Is this the Beginning of the End?

Taleb (2012, 422) remarks "Some theories fall apart, not others" while one famous New York City philosopher Yogi Berra (who some people thought wore a mask and others thought he guarded a national park) allegedly quipped "The future ain't what it used to be." This astute observation was certainly true beginning in September 2006, but does it have to be re-invoked once again at the time of this writing? The central thesis of this book is the price links between crops and biofuels, and between biofuels and energy prices. Events can occur such that these two sets of price linkages can be broken, reducing the possible relevance of the theory presented so far. We now discuss these possible scenarios in turn.

Is the Corn-Ethanol Price Link a Thing of the Past?

We have discussed at great lengths that the excess profits in corn ethanol production since August of 2013 has only returned to zero in September 2014. This has broken the corn-ethanol price link of equation (2.1) for almost the entire time period. This means the actual corn price, for a given observed market price for ethanol, is below that predicted by our theory that assumes (in the long run) zero profits. Although Irwin and Good (2014) write that these excess profits "defy basic economic logic" and are "unsustainable," there can be a set of special circumstances that may be happening now and other events that can occur in the future—shall we say a perfect storm of factors—that can have excess profits for as far as the eye can see. Such factors include another record corn crop in the making and the blend wall becoming less of a concern because the Environmental Protection Agency (EPA; in November 2013) has clawed back, for the first time ever, the renewable fuel standard (RFS) for 2014. Throw in a "collapse" of the crude oil price to, say, $80 per barrel,[5] combined with a continued decline in fuel consumption (e.g., because of efficiencies that increase miles traveled per gallon and a continuing decline in the fuel demand elasticity),[6] and all of a sudden we may be in the throes of the former business as usual in an era where direct crop–biofuel price links did not exist. Or would we?

Let us discuss the outcome of such a possible scenario. Our theory assumes zero profits in the long run. Being able to explain the prediction errors in Figure 2.2 for corn-ethanol and Figure 9.3 for US soybean oil- and EU rapeseed oil-biodiesel gives credence to such an assumption. However, the extraordinary profits in 2013–2014 for US corn-ethanol and EU rapeseed-biodiesel production have not gone away. In the US corn-ethanol case, it reflects in part the reluctance of the US industry to expand ethanol production capacity.[7] There are many reasons for this reluctance and the behavior of the ethanol industry may, in fact, be rational.

First, robust export markets (for ethanol at the time of writing) may not be relied upon to justify ethanol production capacity expansion because world biofuel consumption has leveled off (see Figure 1.1) even before the US mandate reduction in November 2013. The European Union has wavered on their original first-generation mandates, scaling them back and not strictly enforcing them. With suddenly lower ethanol prices, there are now reports that the EU industry is considering exporting ethanol to Brazil[8] whereas Spain is exporting ethanol to Puerto Rico (a US territory).[9]

Moreover, other countries have achieved their more modest (compared to the United States, at least) biofuel targets.

Second, export markets have become more unreliable. The import duties imposed by the European Union on US biodiesel and ethanol exports and on Argentina's biodiesel exports have reduced world prices and diverted biodiesel to the United States, which is now a net importer of biodiesel. The US still exports ethanol to the European Union through third-country loopholes (e.g., through Norway),[10] but it is uncertain how long that can be maintained.

Third, we showed that biodiesel "overage" (the quantity over the biodiesel mandate) multiplied by 1.5 (a conversion factor used by the EPA to convert biodiesel into a corn-ethanol equivalent) and other biomass-based diesel (like renewable diesel) have been primarily responsible for overcoming the ethanol blend wall in 2013 and 2014. In that time period, the amount of ethanol blended in E85 has increased insignificantly. Given that renewable diesel relies on vegetable oils as an input, biomass-based diesel will keep pressure on crop market prices. However, the pressure will be smaller than with ethanol because of the 1.5 equivalence to ethanol for the purpose RFS and because, as we showed earlier, the impact of biodiesel prices on crop prices is lower than ethanol's impact on corn prices.

Fourth, there are other biofuels to overcome the blend wall like bio-gasoline and bio-butanol. Furthermore, cellulosic biofuel production is beginning to take hold and will displace biodiesel in the advanced mandate [receiving higher renewable identification number (RIN) prices should the EPA begin mandating cellulosic ethanol]. This will result in more biodiesel competing directly with ethanol in the Renewable Fuel Gap, and lower D4 and D6 RIN prices, thereby further reducing the need for corn-ethanol to overcome the ethanol blend wall.

But will corn and ethanol prices still be linked through the zero profit condition of equation (2.1) in Chapter 2? This seems less plausible for the future because corn-ethanol production capacity until now was encouraged by the mandate. However, as we explained in Chapter 10, there is no corn-ethanol mandate per se; corn-ethanol is only eligible for the residual portion of the RFS (what we called the renewable fuel gap) for which any biofuel is eligible: sugarcane ethanol and biomass-based diesel (the latter with multipliers of 1.5 and higher). Moreover, this residual mandate maxes out in 2015 at 15 billion gallons. Although corn-ethanol production capacity has increased in tandem with this residual mandate over the years [allowing the zero-profit condition and equation (2.1) to be a fairly accurate representation of the market], any expansion of corn-ethanol production capacity from here on in will not have the safety of a mandate to back it up should market conditions change.[11]

This makes us less sure that corn and ethanol prices will be locked through the zero-profit condition in corn-ethanol production unless there is excess capacity or close to it. As we showed in Figure 2.2, any ethanol production capacity constraints in the past were overcome in due course, but not since August 2013 except for September 2014. However, the future is fraught with many uncertainties, and given the possibility of a "perfect storm" of events described above having implications for our "theory of biofuel policy," in other words, it may well be the case that the corn–ethanol price link is broken, or at least, not expected to hold like in the past. One could imagine a volatile profit premium due to risk and uncertainty.

Can Ethanol Prices Float Down and Away from Gasoline Prices?

The ethanol–gasoline price link is the other key price link in our theory of biofuel policies and their impact on grain/oilseed prices. We showed this price link is expected to be broken in that, with mandate premiums over the tax credit, ethanol prices *float up and away* from gasoline prices. The lowest that ethanol prices can go was given by equation (2.3). However, our analysis of mandate premiums over the tax credit (in Figure 3.2) showed several short periods of negative mandate premiums, contrary to the theory of rational consumers and zero profits in the blending industry. The reason was a series of short-term blending capacity constraints (the last one occurring in the first half of 2010). However, the ethanol blend wall in the future can cause such negative mandate premiums, where ethanol prices *float down and away* from gasoline prices.

Again, even with zero profits in ethanol production, our predicted corn prices would be lower than observed corn prices.

So far, the ethanol blend wall has not caused corn prices to collapse as there have been ethanol mandate price premiums (see Figure 3.2). Furthermore, the D6 RIN price has been a tax on E10 to subsidize the consumption of E85. This has maintained ethanol prices at or above their miles equivalent value relative to gasoline.[12] But imagine our own "perfect storm" of coinciding events discussed above taking place in the future: record corn supplies year after year, reduced motor gasoline consumption, a blend wall, inadequate export markets, biomass-diesel overage due to the structure of and incentives given in the current US biofuel mandate, and increasing amounts of cellulosic ethanol with higher RIN prices, but this time higher crude oil and gasoline prices. This could cause a scenario to play out in the United States like that in Brazil where, for the last eight years in the first decade of the 2000s, E100 was sold at an average discount to E25 of 20 percent (on a miles traveled basis)—see Figure 11.1. This was to encourage the purchase of flex cars and E100 consumption (in other words, to overcome the blend wall that existed in Brazil for many years).

The reason is that the mandate premium can go to zero and the ethanol price is locked onto gasoline price but at a discount to overcome the blend wall. The price paid by blenders for ethanol will be the same for both E10 and E85 with low D6 RIN prices (recall, the scenario has record corn supplies and high crude oil/gasoline prices). Using current US gasoline prices, a 20-percent discount would amount to a $1 per gallon discount in ethanol prices.[13] However, this discount would apply for all corn- (and sugarcane-) ethanol blended into E10 and E85 (as was the case in Brazil).[14] Will the United States enter the nuclear winter Brazil's ethanol industry faced in the past? Only time will tell.

Nevertheless, the implication for predicted corn prices can be enormous. The corn-ethanol price multiplier has averaged approximately four and, therefore, a $1 per gallon ethanol price discount translates into an over $4 per bushel decline in corn prices. This means corn prices are $4 per bushel lower than what our model predicts (and that assume zero profits in ethanol production). Given near futures corn prices of $3.28 per bushel at the time of this writing, we do not predict negative corn prices! But recall Table 9.2, where $135 ($150) per barrel crude oil prices would have corn prices at $6 ($7) per bushel (the lowest corn prices can go, assuming no profits and equation (2.3 to hold). This leaves room for a $4 per bushel price cut due to the blend wall (to see whether it would be the full $4 per bushel, would require further analysis).

To summarize, there are several scenarios under which either price linkage (or both) can be broken by events in the future. However, this does not change the proposition that our theory is most relevant in explaining food grain and oilseed price levels and volatility from September 2006 until now—all other models that have failed to integrate our framework in explaining food commodity prices have, in essence, "fallen apart." But we, nonetheless, heed Yogi Berra's wise counsel about the future (as our profession should have done in September 2006). Or putting the famous New York City philosopher's dictum into more plain language: what is past may not be future. And Taleb's insight should also be kept in mind—the events that required the development of our theory so central to explaining high and volatile grain and oilseed prices in the biofuel policy era came in with a bang. Could they disappear just as quickly?

The Way Forward

There is no simple way to summarize this book. It is more a methodology than results. Implications are situation-specific and as we saw all along from September 2006 to now, there was always something different that caused prices to rise or fall, or different reasons for capacity constraints and prediction errors. We presented models on three basic feedstocks (corn, soybean/rapeseed oil, and sugarcane) and highlight the very unique economic relationships with co-products versus joint products versus competing products. Each biofuel feedstock has a unique economic impact for a given change in the biofuel price, but through the law of one relative international price, markets adjust and relative grains and oilseed prices essentially remain unchanged. The absolute level of these prices varies, depending on the policy, the country and the feedstock despite the stability of relative prices. Some feedstocks have less of an impact for a given biofuel price change than others.

With the focus on price linkages, we showed in Chapter 3 how low corn prices can go and found that 80 percent of the corn-price increase from 2006–2007 to 2011–2012 would have occurred regardless of supply/demand shifts and changes in inventory levels. This is because, given the new corn-ethanol price links, this is the lowest corn prices could have gone. This is the new reality and hence new counterfactual. We showed the biofuel policies caused another $1 billion per day in subsidies to crop farmers, except this time it is a "real" subsidy, raising prices higher than they would otherwise be. However, nobody talks about it because nobody knows. We showed the varied sources of increased price volatility due to biofuels policies, not just crude oil price shocks when the tax credit is binding or crop supply shocks when the mandate is binding (and regime changes) but complex interactions between policies, biofuels, commodities, and countries (not to mention the complex interaction effects with the structure of the US mandate alone or the effects of capacity constraints in ethanol production, marketing, and blending on price volatility). Therefore, many papers have been written on "coping with price volatility" or "variability" since 2008, but how can one make policy prescriptions without understanding the sources of variability?

Our forensic analysis in Chapter 4, for example, showed that the corn-price peak of June 2008 was not due to Midwest flooding as Abbott et al. (2009) claim it to be (or that corn prices fell afterwards due to improving supply conditions). Nor was the level of stocks a determining factor of corn prices at that time, although Wright (2014, 93) asserts "The most important analytical element in this paper that I have added to…Abbott, Hurt, and Tyner…is the discussion of the dynamic response of stocks…"

The law of one international relative price was a key determinant of wheat and rice prices as all other major crop prices were locked onto crude oil prices. We showed that the interdependence of crop prices increase due to biofuels as corn oil is extracted from the co-product of corn processed into ethanol and sold as biodiesel and the co-product price varies with wheat prices as both have higher protein content. The 2008 price boom has been the most studied food commodity price booms of all time and at the same time has been the most misdiagnosed one of them all as well.

Some of our conclusions depend heavily on whether biodiesel is more efficient than ethanol on a relative miles traveled basis when compared with diesel versus gasoline, in addition to the relative fuel taxes (and tax exemptions) applied to such fuels. No general conclusions were reached as it is multidimensional and complex; much more research has to be done to be able to apply this to specific cases and contexts.

Although we discussed at great length the interactions across biofuel policies, feedstocks, and countries, much more work has to be done on that alone, before we go far afield and begin discussing extensions of our work. The important extensions required are multi-dimensional: the integration of our theory with traditional models of price determination, the rationalization of the high-frequency time series econometric literature, which basically needs to start again at square one.[15] Other important areas that require added analysis and research include: uncertainty and capacity constraints as future expansion of biofuel production from food crops does not have the luxury of mandates backing them up, the importance of Hotelling's law of exhaustible resources and the implications for biofuels and hence food commodity prices, and technology policies and breaking the food versus fuel trade-off. These are just to name a few extensions. These issues are not easy to solve but are central issues (and there is no room for instant economics here or for shortcuts).

Let us take one example of extending our framework with traditional models: the theory of rational storage. This is a top priority. If corn prices are locked onto crude oil prices, then the demand curve is essentially flat, so what does the competitive theory of storage have to say about flattening the total demand curve anymore? Without any government biofuel policy, crude oil at $150 per barrel translates into $7 per bushel corn prices (assuming no constraints in ethanol production, marketing or blending, and a competitive equilibrium).[16] What do stockholders do? Speculate on oil prices? If so, the empirical outcome would be exactly the same as would be predicted by the pre-September 2006 model of rational storage: a negative relationship between prices and the stock-to-use ratio.

Likewise, the theory developed in this book has to be integrated with the standard model of developing country policy responses. We already argued that the incidence of these policy responses in 2008 (the year pertaining to all studies of developing country policy response) was mostly on lowering the domestic prices with little if any impact on world prices, again because of the crop–biofuel price linkages. However, more work has to be done on this topic alone.

This need for further research extends to all issues we have touched upon. Farm subsidies in the United States have risen substantially with high prices as crop and revenue insurance subsidies have become the biggest subsidy programs. Therefore, the interaction with farm subsidies is another issue that has to be studied in greater detail than that outlined in this book. Our framework of analysis has extremely important implications for technology policy but has not been given any serious thought although policymakers have put their chit on technology as the solution to the so-called food crisis.

Our theory as presented in this book goes a long way to explain why high price levels and volatility has occurred in the new biofuel era. It is not just biofuel policy regime changes due to shocks in either food commodity markets or energy markets but there are also new sources of volatility due to biofuel policies like interactions of biofuel policies across policy instruments, biofuels, and countries. Capacity constraints in ethanol blending, production, and distribution have also had significant short-run impacts on prices.

Therefore, in that sense, the economic framework of analysis developed in this book is in no way the final word but serves as a springboard to be developed further and be integrated into the conventional models of market price analysis. We identify five priorities: further develop the ideas in this book; integrate the current (and the further developed) theory of biofuel policy into conventional models of price determination; revamp the high-frequency time series econometric literature to take into account the realities of price linkages (as in the discussion of Figure 7.1 in Chapter 7); integrate Hotelling's law of exhaustible resources into our framework as crude oil price is a focal point; and integrate the theory of technology policy into our framework.

However, perhaps most importantly, the analysis ultimately has to be extended to do normative policy analysis (what ought governments to do?), which we have not even begun to fathom. Clearly, the agenda is such that we need to be more demanding and, thus, there will be no room for instant economics or short cuts.

NOTES

Preface

1. These papers were subsequently published (in order) in de Gorter and Just (2009a, b; 2008a, b).
2. We originally underestimated the ethanol price premium. It is now estimated to be $3.30 per bushel in 2006–2007 (higher than the corn price itself for that year!) using more relevant price data and updated multipliers and other parameters.
3. This paper was subsequently leaked to the press and so was published in Mitchell (2008).
4. This simply means the supply curve for ethanol did not intersect (was higher than) the demand curve for ethanol; there was a gap that had to be made up before these two curves would intersect. Thus, there was some "redundancy" in the calculated ethanol price premium due to biofuel policies that really did not impact corn prices; otherwise, corn prices would have been much higher as the price premium we had originally calculated for 2006–2007 was almost equal to the actual corn price!
5. We argue the free market would not have established this price link and profitability in ethanol production in such a short time period could not have generated a market-driven corn price increase in 2008. Perhaps after this first price peak free-market ethanol production would have occurred for the next price peak in 2010–2011.
6. Pinstrup-Andersen (2013a) notes there had been a 75-percent decrease in real grain prices from the 1970s to the end of the twentieth century, so there was a need for an adjustment (the correction in 1996 did not amount to much).

Introduction

1. Many economists ignore the fact that stocks are depleted because of the unexpected impact of biofuel policies.
2. We categorize four alternative literatures apart from our own in Chapter 5 and review these various literatures in Chapters 5–8.
3. As we will show in a moment, the de facto ban of MTBE, a petroleum-based competitor to ethanol, along with the 1978 tax credit, the latter activated by high oil prices in the second-half of the first decade of this century,

were the key environmental and energy policy, respectively, that precipitated the food grain/oilseed price links to ethanol and the ensuing increase in crop price levels and volatility.

1 How Biofuel Policies Ushered in the New Era of High and Volatile Grain and Oilseed Prices

1. Provided there are no capacity constraints in the production, marketing and blending of ethanol, and hence no delinking between corn and ethanol prices.
2. The trend line in Figure 1.3 is for data through 2005 only.
3. Baffes and Dennis (2013) find that the price transmission elasticity of crude oil to grain prices is 20 percent, with the effect being through increased input prices.
4. We define a "mandate" as being a situation where consumers have no choice to choose ethanol on the basis of miles traveled. This "no choice" and hence ethanol price premium above that generated by the tax credit alone can be a result of several situations, including the formal mandate itself; de facto mandates in the form of environmental regulations (see Box 1.2 for details); when consumers are unable to choose between ethanol and gasoline at the gasoline pump on a miles obtained basis because of a lack of flex cars and E-85 stations, so ethanol may be sold at a premium; or when prices are determined in world markets where biofuel policies in the rest of the world are determining the ethanol price premium (perhaps with the help of tax credits on exports as was the case in the United States up to 2012).
5. Economists who emphasize quantity to explain the price boom (unlike price links in this book) should at least emphasize the correct quantities like feed wheat demand increasing because of high corn prices due to ethanol policies rather than a one-off episode of bad weather.
6. This assumes competitive markets, no abnormal profits, and, therefore, no biofuel capacity constraints in the production, marketing, or blending of ethanol.

2 The Economics of Biofuel Policies: The Theory of Corn–Ethanol and Ethanol–Gasoline Price Links

1. This chapter is based on Drabik (2011) who generalizes the pioneering work of de Gorter and Just (2009a, b), Cui et al. (2011), and Lapan and Moschini (2012).
2. The soybean oil–biodiesel model is developed in Chapter 9 and the Brazilian cane-ethanol (made from sugarcane) and sugar complex is outlined in Chapter 11. The biodiesel and Brazil cane-ethanol markets represent the other two major basic feedstocks in biofuel production and are different not only because of the unique structural characteristics of the product but also because of the differing biofuel policies used across regions.

3. The consumption subsidy in the United States takes the form of a blender's tax credit whereas, in the rest of the world, it is implemented as a fuel excise tax exemption.
4. Another issue is that biofuels from some countries do not count toward the domestic mandate because they do not meet the sustainability thresholds. However, it is this threshold (or tariffs, for instance) that discriminates against trade, not the mandate or the tax credit.
5. http://www.card.iastate.edu/research/bio/tools/hist_eth_gm.aspx
6. See de Gorter and Just (2008b, 2009b) and Mallory et al. (2012) for the justification of the zero-profit condition.
7. This formula was originally presented by de Gorter and Just (2008b) and later confirmed theoretically and empirically by Drabik (2011), theoretically by Cui et al. (2011) and Lapan and Moschini (2012), and empirically by Mallory et al. (2012) using futures prices instead.
8. Using cash prices, Abbott et al. (2008, 2009) argue the multiplier is less than two [with no justification as to why it differs from that of de Gorter and Just (2007d, 2008b)]; Mallory et al. (2012) argue the multiplier is equal to 2 when using cash prices while Babcock (2012a) obtains a value of 2.9. The crop−biofuel price link holds regardless of what biofuel policy is binding and only diverges if there are market imperfections like ethanol production or marketing capacity constraints.
9. It is theoretically possible that the relative price of DDGS increases if the demand by hog and dairy producers determines the DDGS price on the margin.
10. The effect of corn prices on crude oil prices can only occur if a change in ethanol-production affects crude oil prices, an impact to be modest at best, if not indistinguishable empirically.
11. Using average EPA data, de Gorter and Just (2008b) take into account the difference in comparing ethanol and gasoline on the basis of miles traveled per gallon of each fuel, rather than by the energy content of the two fuels. This yields a value of $\lambda = 0.7$. If one simply uses the differential energy content, then the value of λ is 0.66 (=75,700 Btu/115,000 Btu; Btu—British thermal unit) (http://bioenergy.ornl.gov/papers/misc/energy_conv.html). Most of the literature uses the latter value.
12. Interestingly, the price consumers pay for ethanol can be greater or less than the market price for gasoline P_G, depending on the relative values of the gasoline price, P_G, fuel tax, t, and the relative miles per gallon parameter, λ.
13. Another penalty often imposed on biofuels is when the consumer price of gasoline is fixed below market prices or is subsidized; thereby causing a reduction in the price that refiners are willing to bid for the biofuel (e.g., many EU countries subsidize diesel consumption).
14. Interestingly, Brazil's tax exemption for ethanol is often not very much higher than the volumetric tax penalty (see Kliauga et al. 2011).
15. The market price of ethanol can be above or below the consumer price paid for ethanol $\lambda(P_G + t)$, depending on whether the tax credit t_c is greater than

or less than the fuel tax t. For a more detailed derivation of equation (2.3), see de Gorter and Just (2008b).

16. A complete exposition of the full theory—allowing for an endogenous crude oil (gasoline) price—is given in de Gorter and Just (2009b).
17. The data values for September 2014 in Figures 2.2, 2.3, and 2.4 are near futures prices as of September 18, 2014.
18. Contrast our analysis with that of Abbott (2014) who identifies approximately 10 different "watershed" time periods between 2007 and 2011 in which capacity constraints play a central role affecting the corn−ethanol price relationship, unlike our analysis summarized in Figure 2.2 which only identifies three episodes in that specific time period. To give an idea of our different approaches, Abbott (2014) specifically identifies June 2009 through December 2009 as the "commodity boom" era. We identify a disequilibrium for September 2009 through February 2010 where capacity constraints arise because of the fall out of an unusual number of plant bankruptcies (a "bust", not a boom being the causal factor).
19. For the various reasons for abnormal profits, see Irwin and Good (2014).
20. The first (and last) corn-ethanol plant to break ground in five years began in August 2013 for 65 million gallons per year capacity (0.44 percent of total US capacity). http://www.ethanolproducer.com/articles/10129/dakota-spirit-breaks-ground-on-65-mmgy-corn-ethanol-plant-in-n-d
21. Intuitively, the minimum price that the producer requires to start ethanol production is higher than what the consumer is willing to pay given that conventional gasoline is available and yields more miles per gallon than ethanol. This concept has, perhaps, confused readers over the years but it is a very basic one: think of the (old) textbook example where the demand curve for televisions on refrigerator doors does not intersect the supply curve (which by the way is not a reality any longer in this modern era!).
22. The only known model of ethanol policy prior to our work in 2007 was Gardner (2007), which ignored the vertical price and horizontal quantity links and had a different and far more rudimentary method for modeling (separately) the tax credit and mandate. Gardner (2007) did not analyze the free market ethanol price, the ethanol price under the tax credit, did not recognize the concept of "water" in the price premium and did not derive the formula for the corn-ethanol price multiplier; instead, he used a value of 2.5 while it peaked at 4.82 in December 2013.
23. The soybean oil−biodiesel model differs with soybean meal and soybean oil as *joint products* and the sugarcane-ethanol model differs where sugar and ethanol are *competing products* from the production of non-traded sugarcane and there are two *by-products*: molasses from sugar used to make ethanol and bagasse for electricity.
24. This link also holds in a special (and rare) case when biofuels are produced without any policy when gasoline prices are sufficiently high and/or the fuel tax sufficiently low.
25. We delay our survey of this literature until Chapter 5.

3 Measures of Biofuel Policy Impact on Food Commodity Prices

1. To avoid the "discontinuities" along the vertical axis in the second panel of the figures (that occur because the conversion factor between ethanol and corn prices is higher than one), and to simplify, we depict the corn market price equal to ethanol prices although, in reality, appropriate conversions have to be made.
2. These shifts can be, for example, due to a depreciating US dollar exchange rate, bad weather, income growth in developing countries, or any other of the perfect storm factors most economists attribute to the food commodity price increase (Heady and Fan 2010; Abbott et al. 2008, 2009; Hochman et al. 2011).
3. We write "or" because, as we show in Chapter 12, the tax credit and mandate price premiums are not additive. If the tax credit is binding, then the mandate is redundant, but not vice versa. This will be explained in Chapter 12.
4. At this stage, we aim at determining the amount of yellow corn to be used in ethanol production at price P_E. When ethanol is produced and the co-product returned in the corn market, then D_{NE} represents demand for corn equivalent, and the implicit demand for yellow corn for non-ethanol use is derived.
5. In reality, DDGS does not necessarily replace yellow corn one-to-one, as the market price of DDGS is not always equal to the price of corn. We capture this by considering the relative price of DDGS and corn in equations (2.1) and (2.1′).
6. For example, Babcock (2013) exactly re-derives equation (2.1) independently of the original analysis in de Gorter and Just (2007d; 2008b) and, therefore, correctly derives the "vertical link" in price space but models the "horizontal link" in quantity space incorrectly by following the approach by Gardner (2007).
7. The ethanol price is adjusted for a decline in the gasoline price due to ethanol production; although modest, it reduces the impact of ethanol policy on corn prices (see Drabik 2011, for details).
8. The premium over the tax credit does not allow for "negative mandate premiums" (negative values reflect ethanol blending-capacity constraints) and predicted corn prices with the tax credit are not allowed to fall below Drabik's (2011) estimates of the non-ethanol corn price P_{NE}.
9. Profits average eight cents per gallon or 23 cents per bushel for the time period under investigation.
10. This was the average value in 2010–2011.
11. This represents 80 percent of the total $519 billion loss in consumer surplus due to the price change reported in Roberts and Schlenker (2013).
12. The title of Peterson's (2009) excellent textbook on the economics of farm subsidies, with more of an international perspective than most such textbooks, is "A Billion Dollars a Day".

13. http://www.engineeringnews.co.za/article/farm-subsidies-and-globetrotting-cows-2002–10–18
14. http://www.ncbi.nlm.nih.gov/pmc/articles/PMC1326363/
15. http://www.nytimes.com/2002/09/30/world/rich-nations-are-criticized-for-enforcing-trade-barriers.html
16. This *law* is explicitly assumed in Roberts and Schlenker (2013) where they add the calories-weighted quantity of corn, wheat, rice, and soybeans into one commodity.
17. Using data from Ivanic and Martin (2014b), consumption weighted international food commodity prices exceeded domestic prices in developing countries by 28 percent in 2007–2008 while by only 9 percent over the entire time period.
18. DTN Daily Ethanol Comments, July 26, 2013. http://www.dtnprogressivefarmer.com/dtnag/common/link.do;jsessionid=32783B6D5252190676CB26C4FD021F60.agfreejvm1?symbolicName=RENEWABLE_FUELS_PAGE_FREE
19. See Irwin and Good (2014) for an extensive list of reasons for the existence of excess profits.
20. We ignore issues of the blend wall which will be discussed in a later chapter.
21. We delay our discussion of how corn prices could go down even further if the blend wall and market conditions were such that the ethanol price was at a 20-percent discount to induce E85 sales.
22. The average value of the corn-ethanol multiplier over the same time period is given by the first term on the right hand side of equation (2.1) in Chapter 2.
23. This does not adjust for downward pressure on gasoline prices with more ethanol production and assumes a long run equilibrium where there would have been no uncertainty as to the tax credit lasting and so investments in ethanol production plants are assumed to have taken place.
24. We use the same method here as we did in Table 3.4.
25. Tyner (2012, 50) lumps de Gorter and Just (2010a) with other studies arguing the tax credit is redundant. This is patently false as the papers by de Gorter and Just always argued the tax credit was never redundant as (1) the tax credit subsidizes fuel (mostly gasoline) consumption when the mandate is binding (de Gorter and Just 2009b); and (2) when the mandate premium is zero, the tax credit increases the ethanol and hence corn price by the full amount of the tax credit, adjusted for changes in gasoline prices due to *changes* in ethanol production resulting from the tax credit (de Gorter and Just (2008b; 2009a; Drabik 2011).

4 A Forensic Analysis of the Food Commodity Price Boom of 2008

1. Meanwhile, soybean- and rapeseed-oil prices were tightly linked to biodiesel prices in the United States and European Union, respectively. But we delay our discussion on biodiesel until Chapter 9.

2. We add the word "traditional" to the Timmer and Dawe (2010) quotation as a new economics of crop prices has emerged, anchored in energy and environmental legislation that developed unique relationships between energy, biofuel and crop prices (as described in this book).
3. Recall that a "mandate" means situations beyond the US blend mandate requirement and includes all situations restricting consumer choice like de facto mandates in the form of environmental regulations or the transactions costs at gasoline retail outlets preventing price competition between say E5 and E10, or when domestic ethanol prices are determined by world prices.
4. For a detailed theoretical explanation of mandates versus tax credits, see de Gorter and Just (2009a, b).
5. At high oil prices in 2008 (and again in 2011), the ethanol price premium due to the mandate fell to zero—the tax credit was binding and corn and oilseed prices were firmly locked on to oil prices in these situations.
6. Our central thesis is that if Asian governments, state trading enterprises, and peasants did not make rice prices keep their relative value to other grains, then other market participants would have. Thus, there has been overemphasis on the impacts of government policy responses and state trader behavior in impacting rice prices (de Gorter et al. 2013b).
7. Abbott et al. (2009) complain about the study by FAO (2008a) that they do not distinguish the effects of biofuels versus biofuel policy on the corn price increase. It would be impossible for the free market to have established that corn-ethanol price link in such a short time when crude oil prices spiked such that in theory, biofuels would have been produced without policy.
8. An exception is the time series econometric literature that seeks a relationship among crop, biofuel, and energy prices using daily, weekly, or monthly data. We critique this literature in Chapter 7.

5 A Critique of the Literature Analyzing Biofuel Policy and the 2008 Food Commodity Price Boom

1. As we shall see in the discussion of the econometrics literature to follow in this chapter and in Chapters 6 and 7, NO study makes this adjustment so in no way do we wish to single Avalos (2014) out here.
2. The introduction to Foucault's (2002) *The Order of Things* quotes from a Chinese encyclopedia where the animals are divided in the following ways: "(a) belonging to the Emperor, (b) embalmed, (c) tame, (d) sucking pigs, (e) sirens, (f) fabulous, (g) stray dogs, (h) included in the present classification, (i) frenzied, (j) innumerable, (k) drawn with a very fine camelhair brush, (l) et cetera, (m) having just broken the water pitcher, (n) that from a long way off look like flies." The chaotic, arbitrary, and artificial characteristics of order, that the example of the Chinese encyclopedia shows so clearly, can be found again exactly in the way the literature classifies the impact of biofuels. There apparently is no principal difference between chaos and order. Only

those who know (or learn) the principles of order, as laid out in this book, will be able to find their way through the labyrinth.

3. Two important points: we write *biofuel policies* because there would be little biofuel production without them (except for Brazilian ethanol under particular market circumstances); and the "correct" economic model has to be used for all issues: analyzing the impact on food commodity prices, the environmental effects, the effects on indirect land use change, and so on.
4. The model we are espousing in this book can be described as one that uses standard economic theories of profit maximization and competitive markets together with basic knowledge of the physical production process for biofuels to examine the potential equilibrium relationships between each market and how these interact with each other [in other words, an explicit focus on the crop (food grain and oilseeds), biofuel (ethanol and biodiesel), energy (gasoline and diesel) and crude oil prices]. We look closely at the biofuel policy that is binding and the specific relations between the various prices. We generally find that biofuels policies play a much bigger role in explaining the recent commodity price increase, and that this role can vary depending on market circumstances.
5. Of course, there is some overlap between literatures (c) and (d) that will become clear once we review literature (d) in Chapter 7.
6. Others, like Dawe (2009), argue low stocks did not have a major impact on rice prices, indicating other factors were at play, such as developing country policy responses, instigated by broader forces like biofuels.
7. See Table 1 in Trostle (2008) and Trostle et al. (2011) for a summary of all the factors considered by the "perfect storm" literature that we summarize in Table 5.1.
8. It is regretful that the Abbott et al. studies did not adopt our approach as Tyner (2008) recognized the new price links between agriculture and energy.
9. We add the gasoline price specifically to question the importance of the perfect storm explanation. Rather, we believe the biofuels effect provides the simplest and most complete explanation. We do not expect corn and gasoline prices to go side by side. Our discussion of Figure 4.4 shows that mandate premiums were common and only when the tax credit is binding would we expect a close connection between gasoline and corn prices.
10. Although Table 3 in Trostle et al. (2011) lists an entire page of policy responses, they select only a few for their Figure 12.
11. There have been many studies done by Iowa State University and the Center for Agriculture and Rural Development (CARD) associated with the names of Bruce Babcock, Dermot Hayes, and others (but we separate the work done at Iowa by Harvey Lapan and GianCarlo Moschini and their co-authors as a genre that undergirds the framework we develop in this book). These studies by CARD all attribute a small effect of biofuel policies on food grain and oilseed prices (e.g., Elobeid and Tokgoz 2008; Babcock and Fabiosa 2011; Babcock 2011).

12. These theories of commodity price determination are based on Dornbusch's (1976) famous theory of exchange rates but with the price of commodities substituted for the exchange rate and with convenience yield substituted for the foreign interest rate.
13. The origins of this approach are found in de Gorter and Just (2008a, b; 2009a, b) and extended, generalized and empirically verified in Drabik (2011). See also Cui et al (2011); Yano et al. (2010, 2011); Collins (2008); de Gorter (2008a, b); de Gorter and Just (2010b); de Gorter and Drabik (2012a, b); Lapan and Moschini (2012); Chen and Khanna (2013); Mallory et al. (2012); and Babcock (2013). de Gorter and Just (2010a) survey this literature up until the end of 2009.
14. The latter literature argues "water" exists in the price premium due to biofuels policies, implying no biofuels would be produced in a free market situation (see especially de Gorter and Just (2008b) and Drabik (2011) for details).
15. It must be pointed out that the competitive theory of storage can simulate price spikes and so this apparently disproves speculation (see Bobenrieth et al. 2014).
16. See http://www.ers.usda.gov/data-products/chart-gallery/detail.aspx?chartId=37396&ref=collection#.U5dmcxAvB3s

6 The Economics of Developing Country Policy Responses and Biofuel Policies

1. For a detailed summary and analysis of all of the types of developing country policy responses, see Sharma (2011), Jones and Kwiecinski (2010), Konandreas (2012), and Anania (2013).
2. This study is a culmination of a series of papers written by Martin and Anderson over the years on the importance of developing country policy responses in the 2008 price boom. Other empirical studies of developing country responses include Bouet and Laborde (2010), Giordani et al. (2012), and Rutten et al. (2013).
3. Rutten et al. (2013) have "Sit down at the Ball Game" in the title of their paper describing the same phenomenon.
4. We will show in the next chapter that, if the mandate is binding, an increase in crude oil prices should have a weak negative relationship with corn prices, ceteris paribus. This does not imply biofuel policies do not have an effect; quite the contrary, as we show in the next chapter, that is when they have maximal effect!
5. We apologize for the rather macabre metaphor; we actually like birds. It is just that economists continually use the terminology in trying to overturn the Tinbergen Rule, the concept that led to the first Nobel Prize in economics.
6. For example, we have not been able to analyze the implications of preexisting farm and trade policies (category {7(b)} in Table 5.1), but Aksoy

and Hoekman (2010, 2) note that: "The causes of the high international food prices included decades of trade-distorting policies that encouraged inefficient agricultural production in rich countries and recurrent dumping of surpluses on global markets, resulting in underproduction in poor countries, thinner global agricultural markets, more volatility, and lower overall reserve supply capacity."

7 An Enhanced Exposition of the Corn–Ethanol–Energy Price Linkages and Implications for Time Series Analysis of Biofuel Policy Impacts

1. The only exception we are aware of is Rajcaniova et al. (2013).
2. In Chapter 9 on biodiesel, we will compare the soybean oil–biodiesel model with the corn-ethanol model. In Chapter 11, we will compare these two models to the sugarcane-ethanol model and summarize the differences in the three models.
3. Of course, one cannot rule out the Green Paradox as biofuel production can be thought of as a tax on crude oil, thereby by Hotelling's theory of exhaustible resources, has crude oil producers increasing supply (Sinn 2008).
4. These numbers correspond to a baseline ethanol consumption of 10.7 billion gallons.
5. Figure 2.3 showed that ethanol prices went below what the minimum ethanol price should have been if the tax credit was binding in three periods in the past—this necessarily means there were ethanol-blending constraints that are expected when ethanol production was growing exponentially in the past. Note that ethanol blending constraints may also have occurred when the mandate was binding before and after the constraint, but there is no way one can identify this with statistical analysis.
6. This is not the case for sugarcane ethanol in Brazil, as will be explained in Chapter 11.
7. This assumes of course that *changes* in biofuel production due to a shock will have little impact on crude oil prices. Otherwise, the analysis will have to be adjusted. However, even total biofuel production has small impacts on the oil price, especially if one incorporates Hotelling's theory of exhaustible resources.
8. The ethanol price floats up and away from its lowest possible value defined in equation (2.2) when linked to the gasoline (and hence crude oil) price through the tax credit.
9. One oil executive termed this "watering down the scotch"!
10. An increase in oil prices causes gasoline price to increase, ethanol price decline as do corn prices (Pathway A to B^m to C^m). Ethanol production costs go up, but have two effects in reducing corn prices: a lower ethanol price lowers corn prices (Pathway D^m; and a processing cost increase, c_o causes a decrease in corn prices (Pathway C^m). Corn production costs go up, shifting the corn supply curve up and directly increasing corn prices

(Pathway E^m) but this also results in a higher ethanol price, dragging corn prices further up, as per equation (2.1), thereby a synergy effect.

11. However, this ignores the input cost channel whereby a higher oil price results into a higher corn price, even under the mandate which can be a stronger effect than the negative effect of the oil price through the ethanol price. Therefore, on the net, even under the mandate the observed relationship can be positive.
12. It should be noted that there are now over 100 papers in this literature, some cited in Serra and Zilberman (2013) are not tabulated in the tables (and more coming seemingly every day!).
13. Balcombe and Rapsomanikis (2008), however, fail to recognize that Brazil's gasoline price is disconnected from world oil prices (see Figure 11.3 in Chapter 11), which can significantly affect their results.
14. An exception is the study by Ciaian and Kancs (2011), who develop a multi-commodity model to investigate the responsiveness of prices of select agricultural commodities to a positive crude oil price shock. And they find the responsiveness varies, as one would expect initially although the law of one relative international price would have all crop prices moving together again. Ciaian and Kancs (2011) extend de Gorter and Just's (2008b) model to allow for agricultural commodities other than feedstock and for the indirect input channel through which energy can affect agricultural prices (but this can only show up in the data when the tax credit is binding so regime changes have to be incorporated in such econometric time series analysis).

8 The Impact of Biofuel Policies on Food Commodity Price Volatility

1. The World Bank began a Food price Watch website http://www.worldbank.org/en/topic/poverty/publication/food-price-watch-home and chronicles World Bank initiatives like coordinating activities with UN agencies here http://www.worldbank.org/en/news/press-release/2014/05/29/global-food-price-increase-august-2012
2. *Teff* is an economically important Ethiopian annual cereal grass (*Eragrostis tef syn. E. abyssinica*) grown for its small grain which yields a white flour and as a forage and hay crop.
3. A good number of books have "volatility" in their titles (e.g., Prakash 2011; Chavas et al. 2014; Piot-Lepetit and M'Barek 2011) as well as the 2014 special issue of the *Journal of International Money and Finance*, and literature surveys like that of Araujo et al. (2014).
4. The two primary paths for volatility to cause harm are behavioral in nature. Volatility could cause harm if short term price spikes are interpreted as long term in nature and lead farmers to make long term plans based on false expectations. Given the staying power of recent high prices, this does not appear to apply. Alternatively, price volatility could cause problems if they lead farmers who are risk averse to cut production, leading to higher average prices overall.

5. In the survey by Brümmer et al. (2013), biofuels is the sixth most common cited driver of price volatility, after financialization and speculation, crude oil prices, low stocks, exchange rates, and increased consumption.
6. Always remember that when we write "mandate," it may refer to the formal mandate, a de facto mandate due to environmental regulations, irreversible investments by refiners using ethanol as an octane enhancer, as well the ethanol blend wall or no choice as transactions costs and constantly changing blend requirements allow blenders to "water down the scotch." The mandate premium may also be due to international trade and prices.
7. Simulations in Drabik (2011) show the change in corn prices due to a corn supply shock are 16 times higher under a mandate than a tax credit. This shows that there is no one-to-one relationship between the effects of shocks on price levels versus price volatility.
8. A thorough exposition of the theory of competitive storage is summarized in Williams and Wright (1991). However, Deaton and Laroque (1992), among others, question the relevance of this theory. Many studies have argued the theory has been rescued (e.g., Cafiero et al. 2011), but we take the relevance of this theory as given. Our objective here is to analyze its significance in this new biofuel era and what adjustments have to be made to make it more relevant.
9. Therefore, economists have understated ethanol's price impact on corn.
10. Our discussion in Chapter 4 had a different reason why prices rose in 2007–2008 and why they did not rise as much in 2009–2010 as they did in 2007–2008 or in 2010–2011.
11. Ironically, the Roberts and Tran (2012) estimate of the corn price is equal to that of Roberts and Schlenker (2013) who did not include inventory behavior in their model.
12. The same pattern exists for corn stocks and prices; wheat prices and stocks went up in tandem only in 2008–2009.
13. Data and parameters for calibrating the 2005–2006 and 2010–2011 crop year simulations came from multiple sources and are documented in Byrd (2012).

9 The Economics of Biodiesel and the Central Role of the European Union's Policies

1. http://stats.oecd.org/Index.aspx?DataSetCode=HIGH_AGLINK_2011
2. http://www.ufop.de/english/news/environmental-committee-rules-in-favour-of-decline-in-european-biofuel-production/
3. http://www.biofuelsdigest.com/bdigest/2013/02/25/spain-lowers-biodiesel-consumption-targets-to-4-1/
4. Some EU countries also set biodiesel specific blending targets. The future levels of EU blending targets have been discussed intensively recently. The EU Commission proposed in 2012 to cap the share of food-based biofuels to 5 percent by 2020. In September 2013, the EU Parliament voted to loosen

the cap to 6 percent and set a separate 2.5-percent target to incentivize production of "second-generation" biofuels made from waste products. But on October 17, 2013 the proposal to cap EU's use of food-based biofuels was stalled by a vote in the European Parliament's environment committee.

5. This ratio does vary, reaching a peak of 2.24 in December 2011.
6. For an explanation for why these sustainability standards make little economic sense, see de Gorter and Just (2009c).
7. The ratio of the weight of world oilseed oil to meals increased from 53 percent in 2005 to 61 percent in 2014 (WASDE, several issues), reflecting the worldwide switch to more oil intensive oilseed crops.
8. In the corn-ethanol case, we have a *co-product* in the form of DDGS while in the sugarcane-ethanol market, we have *competing products* (ethanol versus sugar) and *by-products:* molasses from sugar production used in ethanol production; and bagasse used for electricity. Note: a by-product is not the same as a co-product; the latter refers to a product that is returned to the same market where the original product comes from (e.g., DDGS comes from corn).
9. This section draws from de Gorter et al. (2011).
10. http://gain.fas.usda.gov/Recent%20GAIN%20Publications/Biofuels%20Annual_The%20Hague_EU-27_8-13-2013.pdf
11. de Gorter et al. (2011) provide details on how that happened.
12. In Chapter 12, we highlight the tax credit acts differently with exports than if no trade; and although a blender's tax credit is identical to a tax exemption in a closed economy analysis, it differs in a trade context.
13. Without loss of generality, in the rest of the chapter we assume human consumption is the only non-biodiesel use of soybean oil.
14. There are by-products from biodiesel production (e.g., glycerin) but their value is small and declining (even negative in Europe when the supply of glycerin is too high); hence, we incorporate them into the value of c_{ob}.
15. Lapan and Moschini (2012) present an analogous equation for ethanol and corn prices.
16. The underlying data come from the "Historical Biodiesel Operating Margins" spreadsheet of the CARD of the Iowa State University. http://www.card.iastate.edu/research/bio/tools/hist_bio_gm.aspx
17. See the next chapter and Shepherd (2014) for a detailed explanation of this over prediction.
18. The data on biodiesel processing cost in the European Union (Germany) are not publically available. In order to calculate the predicted rapeseed oil prices, we estimated the processing cost as follows. First, we averaged the biodiesel processing cost reported for a representative plant in Iowa for the period April 2007–March 2014 (http://www.card.iastate.edu/research/bio/tools/hist_bio_gm.aspx). The average was taken because, by assumption (supported also by observed data), this cost is constant. Second, because the natural gas price is the main part of this cost, we multiply the calculated average value of the cost for Iowa by a ratio of monthly natural gas prices in Germany and Iowa. In this way, we construct monthly estimated for the biodiesel processing cost in Germany.

19. The literature on contingent protection shows that imports fall once an investigation has been initiated.
20. http://www.card.iastate.edu/research/bio/tools/hist_eth_gm.aspx
21. Although biodiesel prices are about twice of the ethanol prices, the results are about marginal changes and so a change in biodiesel prices still has a lower impact on crop prices.
22. The US Gulf price of diesel.
23. Note that German biodiesel prices have been stable since January 2013 at \$6.40 per gallon (almost double of the US prices. This explains high profits in Germany in Figure 9.1).
24. The corresponding production coefficients for oil and meal per bushel of canola are β_1 = 22.98 pounds and β_2 = 37.02 pounds.
25. http://www.hort.purdue.edu/newcrop/afcm/canola.html
26. Msangi and Evans (2013) and Shinoj et al. (2011) point out possible problems with using jatropha, especially the absence of an economically usable joint product, that is, meal which adversely affects the profitability of biodiesel production from jatropha.

10 The Complex Structure of the US Biofuel Mandate: A Handbook

1. In November 2013, the EPA recommended a 1.34-billion gallon decline in the RFS for 2014 but insisted it was not a final decision.
2. We are indebted to Joe Cooper, Rob Johansen, and Seth Meyer for conversations on the issues covered in this chapter. They anticipated these issues (see, for example, Meyer et al. 2012; Meyer and Johansson 2013; Thompson et al. 2010).
3. Belzowski and Green (2013) found diesel cars get approximately 34 percent more miles per gallon than the same cars fuelled by gasoline. Yet the European Union gives diesel 12.5 percent higher energy content in the "energy equivalent" mandate for the two biofuels making up the overall mandate.
4. The RFS is under the Clean Air Act Section 211 (o) as amended by the Energy Independence and Security Act (EISA) of 2007 (EISA 2007).
5. Cellulosic biofuels are renewable fuels derived from any cellulose, hemicellulose, or lignin that is derived from renewable biomass.
6. Most biomass-based diesel production in the United States is made from vegetable oils and animal fats.
7. Biofuels themselves are net zero as recognized by the IPCC (2007) (the GHGs absorbed in growing the crop are emitted in burning the fuel—see de Gorter and Just (2009c) for a comprehensive discussion). But GHGs are emitted in the production and distribution of biofuels from field to fuel tank, as measured by "life-cycle accounting."
8. Corn-ethanol produced from facilities that commenced construction after the date of enactment (December 19, 2007).

9. If it were not for the California Low Carbon Fuel Standard (LCFS), then all Brazilian sugarcane-ethanol and renewable diesel is effectively in the renewable fuel gap. But because of the LCFS, part of Brazilian ethanol and part of renewable diesel consumption will occur regardless and so becomes part of the advanced biofuel mandate rather than all be overage.
10. For 2013, the EIA forecasted 15.51 billion gallons of ethanol equivalent biofuel consumption versus a total EPA mandate of 16.55 billion gallons ethanol equivalent. Clearly if a blender followed the EIA forecast (and thus Tyner's calculation as discussed in the main text) they would blend over one billion gallons less than actually required. While this gap can be made up with other compliance pathways (i.e., stored RINs) it is clear that a deeper interpretation is needed for determining the mandated blend ratio.
11. http://www.investingdaily.com/16337/driving-into-the-blend-wall/
12. The EPA (2013c) does give a nugget of information in number four to clarify things: "total actual volume of renewable fuel required may also vary from that used to set the standards if the projected volumes of gasoline and diesel used in the calculation of the standards differs [*sic*] from the actual volumes of gasoline and diesel produced in the U.S."
13. The EIA forecasts a significant increase in drop-in fuels http://www.eia.gov/forecasts/aeo/MT_licuidfuels.cfm
14. It is ironic that the EPA has scaled back the total RFS for the first time ever in November 2013 for the 2014 calendar year citing issues with the blend wall. When the original EISA (2007) legislation was passed, the EIA forecast was for 20 percent of total motor gasoline to be ethanol. So it is not clear why the blend wall has all of a sudden become an issue.
15. The removal of the ethanol tax credit had no effect on D6 RIN prices (for an explanation of what a tax credit does in the presence of a blend mandate, see de Gorter and Just 2009b), yet the biodiesel tax credit impacts D4 RIN prices as the price of biodiesel is higher than diesel prices.
16. RIN prices are given in $ per ethanol gallons equivalent so the "wet" D4 RIN price is actually 50-percent higher for biodiesel.
17. Note that the D4 RIN price can be higher than the D5 RIN price which can be higher than the D6 RIN price. However, the reverse cannot occur so there is an ordering of RIN prices, given the nested structure of the mandate.
18. Over 500 million gallons of "wet" biodiesel and renewable diesel were imported during 2013.

11 The Economics of Brazil's Sugarcane–Ethanol/Sugar Complex and Policies

*This chapter is sourced primarily from de Gorter et al. (2103f) and Kliauga et al. (2011).

1. The United States was increasing the content of ethanol in gasoline to avoid the ethanol import tariff and, instead, pay a lower gasoline import tariff.

2. Recall in our discussion of Table 9.2 that these estimates are first round or direct impacts, ignoring market adjustments and so are only an indicator of the true effects on each feedstock price.
3. The EU mandate has a similar effect with the mandate specified in energy equivalent terms.
4. Or 4.7 times once you include the 1.5 corn-ethanol equivalence values for biodiesel.

12 The Interaction Effects Among Biofuels, Policies, and Countries

1. This is true also for the US tax credit on corn ethanol. Before it expired at the end of 2011, the US tax credit was paid on a majority of ethanol exports; because the ethanol price was determined in Brazil, the incidence of the US tax credit in a higher domestic ethanol price received by ethanol producers.
2. For simplicity, we assume the gasoline price is fixed.

13 The Impact of Biofuel Policies in the Future: Some Concluding Remarks

1. We have done everything in our power to make biofuels a fad but have failed so far; see the surveys by de Gorter and Just (2010a) and de Gorter et al. (2013d) who take a hard look at the key social costs of biofuel policies.
2. To be fair, Tomek and Kaiser (2014) is the leading textbook on agricultural commodity price analysis (in its fifth edition), and if the events we describe in the next section continue or bear out (where price linkages exist no more), then biofuels have just been a blip on the radar screen and do not belong in a textbook written for the ages. But the other book and survey paper cited focus on 2008 and, therefore, are not excused so easily. Having said that, however, we argue that any serious student of price analysis from September 2006 to this day has to immerse themselves into the economics of biofuels policies as this has been the fulcrum upon which the level and volatility of food grain and oilseed prices have played out on.
3. Babcock (2013), for example, is the latest to at least adopt the exact same corn-ethanol price linkage equation.
4. http://brooklynrail.org/2002/07/express/social-science-social-conscience-remembering-pierre-bourdieu. Bourdieux was a prominent French philosopher, taking a sociology approach, and his book was a culmination of nearly two decades of research.
5. Or think of it this way: the higher the gasoline price, the higher the ethanol price will be to overcome the blend wall and the more likely equation (2.1) will hold.
6. One would think such a development would make the blend wall more binding and so increase the decline in prices of ethanol to meet the mandate.

But an increase in fuel demand would give more room for "watering down the scotch" and, therefore, reduce the constraints of a blend wall and allow for higher corn-ethanol prices, ceteris paribus.

7. In Figure 2.2, our model predicts \$9.39 per bushel corn price in December 2014 (assuming zero profits), when observed corn prices were \$4.59 (a 96 percent error).
8. http://www.platts.com/latest-news/agriculture/london/traders-eye-eu-ethanol-exports-to-brazil-on-multi-26738031
9. http://www.biofuelsdigest.com/bdigest/2014/01/27/weak-ethanol-prices-in-europe-see-exports-picking-up/
10. http://www.biofuelsdigest.com/bdigest/2014/02/03/epure-files-complaint-with-ec-against-us-ethanol-imports-via-norway/
11. In fact, we predict that by the beginning of 2015, there will not be 15 billion gallons of corn-ethanol production capacity (the "residual" mandate which most of assumed would be taken up by corn-ethanol), even with substantial corn-ethanol exports. Others had predicted that US corn-ethanol consumption would be greater than 15 billion gallons by 2015.
12. However, since August 2013 these price premiums that otherwise would increase corn prices have been almost exactly offset by excess profits that decrease corn prices (as per our prediction error in Figure 2.2) except for the data point for September 2014 where our estimate of excess profits are zero.
13. Current retail prices of motor gasoline are about \$3.75 per gallon. A 20-percent discount would amount to 75 cents per gallon. Given that E85 is on average 75 percent ethanol, the required discount in ethanol prices would be \$1 per gallon. Note that Babcock and Pouliet (2013) also get a \$1 price discount requirement to achieve sufficient E85 sales.
14. Not like now where the D6 RIN price taxes E10 to subsidize E85, keeping ethanol prices at or above their miles equivalent value relative to gasoline.
15. Especially as every paper claim their analysis has policy implications!
16. And by the law of one price, all other grain/oilseed prices rise too.

DEFINITIONS

Advanced mandate residual (or "undifferentiated advanced biofuels")—one of three categories under the "advanced biofuels" in the RFS: it is calculated as the residual of the advanced mandate subtracting the mandates for cellulosic ethanol and biodiesel.

Anhydrous ethanol—a type of Brazilian ethanol consumed only in blends with gasoline.

Binding—we use this adjective to denote a biofuel policy that determines a biofuel price in a given time period.

Biodiesel—a vegetable oil- or animal fat-based biofuel produced via a transesterification process. Produced, for example, from soybean oil; biodiesel is blended with diesel.

Biofuel—a liquid fuel produced from biomass that serves as a substitute to conventional fossil fuels (i.e., gasoline and diesel).

Biomass-based diesel—any diesel fuel made from biomass feed stocks, including biodiesel (mono-alkyl esters) and non-ester renewable diesel (e.g., cellulosic diesel). Biomass-based diesel excludes renewable fuel derived from co-processing biomass with a petroleum feedstock. The US lifecycle GHG emissions reduction threshold is 50 percent.

Blender's tax credit—a reimbursement made to the fuel blender at the end of a fiscal year to help incentivize purchase of biofuels and increase blending.

Blend wall—a technical maximum (stipulated by the car manufacturer) for the volumetric share of a biofuel in the final fuel blend.

By-product—bagasse used for electricity or molasses from sugar production used in ethanol production.

California LCFS—the California Low Carbon Fuel Standard is like a market-based cap and trade program for transportation fuels only, where the carbon intensity of all transportation fuels (weighted average GHG emission reductions by fuel) is required to decline 10 percent over time.

Cellulosic ethanol—a biofuel produced from lignocellulose contained in wood, grasses, or the inedible parts of plants.

Competing product—a result of a production process that can result in one or several main products, depending on profitability. A typical example is sugar and ethanol in Brazil.

Co-product—a product that is returned back to the market of its input. A typical example is DDGS produced from corn in the process of ethanol production.

Diesel—the second most represented product of crude oil refining.

Ethanol—an alcohol produced, among others, from corn, wheat, sugarcane, or switch grass; ethanol is mixed with gasoline.

Excise tax exemptions—reductions in the consumer excise tax at the pump level (typically outside of the United States).

Feedstock—a raw material used for biofuel production (e.g., corn, rapeseed, straw).

Free market biofuel price—the hypothetical biofuel price under no biofuel policy; it is determined only by the gasoline (diesel) market price and tax.

Gasoline—the main product of crude oil refining.

Generalized Autoregressive Conditional Heteroskedasticity—a generalization of the Autoregressive Conditional Heteroskedastic process that allows for past conditional variances in the current conditional variance equation. This model is frequently used in the HFT literature on commodity prices.

Great Recession—the period in 2008–2010 where the greatest economic depression occurred worldwide since the Great Depression of the 1930s.

Green Paradox—a situation when biofuels lower fossil fuel prices, thereby acting like an announced expropriation of fossil fuel resources from their owners. The resource owners then accelerate the resource extraction, thus increasing fossil fuel supplies and depressing their prices even further.

High-Frequency Time series (HFT)—econometric analysis using daily, weekly, or monthly data.

Hotelling's theory of exhaustible resources—a finding that the most socially and economically profitable extraction path of an exhaustible (i.e., non-renewable) resource is one along which the price of the resource increases at the rate of interest.

Hydrous ethanol—a type of Brazilian ethanol consumed only in its pure (i.e., 100 percent) form, known as E100. Hydrous ethanol consists of 95 percent ethanol and 5 percent water.

Joint product—a product that is jointly produced from some inputs, is not returned back to the input market, and does not compete with another joint product. A typical example is soybean oil and soybean meal produced from soybeans.

Lifecycle GHG emissions—the aggregate quantity of greenhouse gas emissions (including direct emissions and significant indirect emissions such as significant emissions from land use changes) related to the full fuel lifecycle, including all stages of fuel and feedstock production and distribution, from feedstock generation or extraction through the distribution and delivery and use of the finished fuel to the ultimate consumer, where the mass values for all greenhouse gases are adjusted to account for their relative global warming potential.

Mandate price premium—the difference between the biofuel price under the binding mandate and the free market biofuel price.

Mandate premium over the tax credit—the difference between the mandate premium and the blender's tax credit.

New "financialization" of food commodities—new financial investors including banks, financial services firms, and large-scale institutional investors who encouraged herding behavior to the point where financial investment, rather than market fundaments like supply and demand, increasingly influences prices.

Overage—once the advanced mandate is filled, biofuels begin to count toward the renewable fuel gap (residual of the RFS).

Perfect storm—a notion that the spike in the foodgrain/oilseed commodity prices in 2008 (and later) was due to a host of factors that occurred at the same time (e.g., bad weather, growing world population, rising incomes in developing countries, or a change in the macroeconomic environment).

Recycling effect—a notion that DDGS returned to the corn market replaces yellow corn that can be further processed into ethanol.

Renewable diesel—a petrodiesel-like fuel derived from biological sources that are chemically not esters and, thus, distinct from biodiesel. Renewable diesel can be produced from the same feedstock as biodiesel.

Renewable fuel gap—the residual of the total biofuel mandate, after the advanced mandates have been filled.

Splash and dash—under the American Jobs Creation Act of 2004, a tax credit of $1.00 per gallon was paid for any gallon of biodiesel blended in the United States with oil-based diesel, regardless of its source (domestic or imported) or amount (a typical blend was a "splash" of 0.1 percent diesel and 99.9 percent biodiesel). The resulting blend was then exported ("dashed") to the European Union, where some of the biodiesel had even originated.

Standing up in the stadium effect—the decision of some people in the stadium to stand up to see better is beneficial to them only if a few people do so. If all the people stand up, they can see equally well as before, except now all have to stand. Similarly, the larger the number of countries insulating their domestic markets, the more other countries perceive a need to do likewise, which exacerbates the effect on international prices such that even greater changes in trade barriers are desired by each nation, both exporters and importers.

Triple dipping—a situation where Canada exported all of its biodiesel production, taking advantage of its own biodiesel production subsidy, the US tax credit, and the high prices in the EU due to EU excise tax exemptions.

Water down the scotch—a term allegedly coined by an oil executive to describe the situation where consumers do not have a choice to face competition at the fuel pump for different levels of ethanol mixed with gasoline (because of transactions cost); therefore, they have no choice and a fuel blender takes advantage of consumer's inability to choose a fuel based on miles traveled by pricing the biofuel the same as a gallon of a fossil fuel without adjusting the price for the lower miles achieved.

Water in policy premium—the gap between the intercept of the biofuel supply curve and the (lower) free market biofuel price.

Wet gallons—"corn–ethanol equivalent" gallons converts "wet" (actual) gallons into RINs (e.g., 1 "wet" gallon of biodiesel counts as 1 to biomass-based diesel mandate but 1.5 toward residual of RFS or renewable fuel gap).

REFERENCES

Abbott, Philip C. "Biofuels, Binding Constraints and Agricultural Commodity Price Volatility," Chapter 3. In *The Economics of Food Price Volatility*, edited by Jean-Paul Chavas, David Hummels, and Brian D. Wright. Chicago, IL: University of Chicago Press, October 2014.

Abbott, Philip C., Christopher Hurt, and Wallace E. Tyner. "What's Driving Food Prices?" Farm Foundation Issue Report. Oak Brook, IL: Farm Foundation, 2008.

Abbott, Philip C., Christopher Hurt, and Wallace E. Tyner. "What's Driving Food Prices? March 2009 Update." Issue Report. Oak Brook, IL: Farm Foundation, 2009.

Aksoy, M. Ataman, and Bernard Hoekman (eds). *Food Prices and Rural Poverty*. Washington, DC: The World Bank, 2010.

Alston, Julian M., Matthew A. Andersen, Jennifer S. James, and Philip G. Pardey. *Persistence Pays: U.S. Agricultural Productivity Growth and the Benefits from Public R&D Spending*. New York: Springer, 2010.

Anania, Giovanni. "Agricultural Export Restrictions and the WTO: What Options Do Policy-Makers Have for Promoting Food Security?" ICTSD Programme on Agricultural Trade and Sustainable Development Issue Paper no. 50, November 2013.

Anderson, Soren T. "The Demand for Ethanol as a Gasoline Substitute." *Journal of Environmental Economics and Management* 63, no. 2 (2012): 151–168.

Anderson, Kym, Maros Ivanic, and Will Martin. "Food Price Spikes, Price Insulation, and Poverty," Chapter 8. In *The Economics of Food Price Volatility*, edited by Jean-Paul Chavas, David Hummels, and Brian D. Wright. Chicago, IL: University of Chicago Press, October 2014a.

Anderson, Kym, Maros Ivanic, and Will Martin. "Trade Insulation for Food Does not Decrease Global Poverty." East Asia Forum, March 13, 2014b. http://www.eastasiaforum.org/2014/03/13/trade-insulation-for-food-does-not-decrease-global-poverty/

Araujo Enciso, Rene S., Maria Blanco, Marco Artavia, Fabien Ramos, Francisco Fernández, Benjamin van Doorslaer, Davide Fumagalli, and Andrej Ceglar. "Volatility Modelling: Long-Term Challenges and Policy Implications." Scientific Paper no. 5, ULYSSES Project, July 2014. http://www.fp7-ulysses.eu/

Auld, Douglas. "The Ethanol Trap: Why Policies to Promote Ethanol as Fuel Need Rethinking." C.D. Howe Institute Commentary, no. 268, July 2008.

Avalos, Fernando. "Do Oil Prices Drive Food Prices? The Tale of a Structural Break." *Journal of International Money and Finance* 42 (2014): 253–271.

Babcock, Bruce A. "The Impact of US Biofuel Policies on Agricultural Price Level and Volatility." Issue Paper no. 35. Geneva: International Center for Trade and Sustainable Development, 2011.

Babcock, Bruce A. "Biofuels and Food Prices." Presentation at a conference sponsored by the Agricultural Trade Promotion Center, Chinese Ministry of Agriculture and the International Center for Sustainable Trade Policy, Beijing, China, August 2012a.

Babcock, Bruce A. "Preliminary Assessment of the Impacts of the Drought's Impacts on Crop Prices and Biofuel Production." CARD Policy Brief 12-PB7, July 2012b.

Babcock, Bruce A. "Outlook for Ethanol and Conventional Biofuel RINs in 2013 and 2014." CARD Policy Brief 12-PB 9, December 2012c.

Babcock, Bruce A. "Ethanol without Subsidies: An Oxymoron or the New Reality?" *American Journal of Agricultural Economics* 95, no. 5 (2013): 1317–1324.

Babcock, Bruce A., and Jacinto F. Fabiosa. "The Impact of Ethanol and Ethanol Subsidies on Corn Prices: Revisiting History." CARD Policy Brief 11-PB 5, April 2011.

Babcock, Bruce A., and Sebastien Pouliot. "Price It and They Will Buy: How E85 Can Break the Blend Wall." CARD Policy Brief 13-PB 11, August 2013.

Baffes, John. "Oil Spills on Other Commodities." *Resources Policy* 32, no. 1 (2007): 126–134.

Baffes, John. "A Framework for Analyzing the Interplay among Food, Fuels, and Biofuels." *Global Food Security* 2, no. 2 (2013): 110–116.

Baffes, John, and Allen Dennis. "Long-Term Drivers of Food Prices." Policy Research Working Paper 6455. Washington, DC: The World Bank, 2013.

Baffes, John, and Xiaoli L. Etienne. "Analyzing Food Price Trends in the Context of Engel's Law and the Prebisch-Singer Hypothesis." Paper presented at the International Conference on Food Price Volatility: Causes and Consequences, Rabat, Morocco, February 2014.

Baffes, John, and Tassos Haniotis. "Placing the 2006/08 Commodity Price Boom into Perspective." Policy Research Working Paper 5371. Washington, DC: The World Bank, 2010.

Balcombe, Kelvin. "The Nature and Determinants of Volatility in Agricultural Prices: An Empirical Study." In *Safeguarding Food Security in Volatile Global Markets*, edited by Adam Prakash, 89–110. Rome: FAO, 2011.

Balcombe, Kelvin, and George Rapsomanikis. "Bayesian Estimation and Selection of Nonlinear Vector Error Correction Models: The Case of the Sugar-Ethanol-Oil Nexus in Brazil." *American Journal of Agricultural Economics* 90, no. 3 (2008): 658–668.

Ball, Eldon, David Schimmelpfennig, and Sun Ling Wang. "Is U.S. Agricultural Productivity Growth Slowing?" *Applied Economic Perspectives and Policy* 35, no. 3 (2013): 435–450.

Baltzer, Kenneth. "International to Domestic Price Transmission in Fourteen Developing Countries during the 2007–08 Food Crisis." WIDER Working

Paper no. 2013/031, March 2013. http://www.wider.unu.edu/publications/working-papers/2013/en_GB/wp2013-031/

Barrett, Christopher B. *Food Security and Sociopolitical Stability*. Oxford, UK: Oxford University Press, 2013.

Barrett, Christopher B., and Marc F. Bellemare. "Why Food Price Volatility Doesn't Matter Policymakers Should Focus on Bringing Costs Down." *Foreign Affairs*, July 12, 2011. http://www.foreignaffairs.com/articles/67981/christopher-b-barrett-and-marc-f-bellemare/why-food-price-volatility-doesnt-matter

Baumeister, Christiane, and Lutz Kilian. "Do Oil Price Increases Cause Higher Food Prices?" Presentation at 59th Panel Meeting of Economic Policy, Izmir, Turkey, April 2014.

Bellemare, Marc F. "Rising Food Prices, Food Price Volatility, and Social Unrest." *American Journal of Agricultural Economics* 97, no. 1 (2015): 1–21.

Bellemare, Marc F., Christopher B. Barrett, and David R. Just. "The Welfare Impacts of Commodity Price Volatility: Evidence from Rural Ethiopia." *American Journal of Agricultural Economics* 95, no. 4 (2013): 877–899.

Belzowski, Bruce M., and Paul Green. "Total Cost of Ownership: A Gas Versus Diesel Comparison." University of Michigan Transportation Research Institute, March 2013.

Berazneva, Julia, and David R. Lee. "Explaining the African Food Riots of 2007–2008: An Empirical Analysis." *Food Policy* 39 (2013): 28–39.

Bobenrieth, Eugenio S.A., Juan R.A. Bobenrieth, and Brian D. Wright. "Bubble Troubles? Rational Storage, Mean Reversion and Runs in Commodity Prices," Chapter 5. In *The Economics of Food Price Volatility*, edited by Jean-Paul Chavas, David Hummels, and Brian D. Wright. Chicago, IL: University of Chicago Press, October 2014.

Bouet, Antoine, and David Laborde "Economics of Export Taxation in the Context of Food Crisis: A Theoretical and CGE Approach Contribution." International Food Policy Research Institute, Discussion Paper no. 00994, 2010.

Brümmer, Bernhard, Olaf Korn, Kristina Schlüßler, Tinoush Jamali Jaghdani, and Alberto Saucedo. "Volatility in the After Crisis Period – A Literature Review of Recent Empirical Research." A Working Paper, 2013. http://www.fp7-ulysses.eu/publications.html.

Byrd, Elizabeth S. "Application of the Competitive Storage Model to U.S. Corn Ethanol Policy." Master's Thesis, Charles H. Dyson School of Applied Economics and Management, Cornell University, 2012.

Cafiero, Carlo, Eugenio S.A. Bobenrieth H., Juan R.A. Bobenrieth H., and Brian D. Wright. "The Empirical Relevance of the Competitive Storage Model." *Journal of Econometrics* 162, no. 1 (2011): 44–54.

Carter, Colin, Gordon C. Rausser, and Aaron Smith. "Commodity Booms and Busts." *Annual Review of Resource Economics* 3 (2011): 87–118.

Carter, Colin, Gordon C. Rausser, and Aaron Smith. "The Effect of the U.S. Ethanol Mandate on Corn Prices." A Working Paper, 2013. http://www.ourenergypolicy.org/wp-content/uploads/2013/07/The-Effect-oftheUS-Ethanol-Mandate-on-Corn-Prices-.pdf

Cevika, Serhan, and Tahsin Saadi Sedik. "A Barrel of Oil or a Bottle of Wine: How Do Global Growth Dynamics Affect Commodity Prices?" *Journal of Wine Economics* 9, no. 1 (2014): 34–50.

Chavas, Jean-Paul, David Hummels, and Brian D. Wright (eds). *The Economics of Food Price Volatility*. Chicago, IL: University of Chicago Press, October 2014.

Chen, Xiaoguang, and Madhu Khanna. "Food vs. Fuel: The Effect of Biofuel Policies." *American Journal of Agricultural Economics* 95, no. 2 (2013): 289–295.

Ciaian, Pavel, and d'Artis Kancs. "Interdependencies in the Energy–Bioenergy–Food Price Systems: A Cointegration Analysis." *Resource and Energy Economics* 33, no. 1 (2011): 326–348.

Collins, Keith. "The Role of Biofuels and Other Factors in Increasing Farm and Food Prices." Report prepared for Kraft Foods Global, Inc., Glenview, IL, 2008.

Condon, Nicole, Heather Klemick, and Ann Wolverton. "Impacts of Ethanol Policy on Corn Prices: A Review and Meta-Analysis of Recent Evidence." NCEE Working Paper 2013–05, August 2013.

Cui, Jingbo, Harvey Lapan, GianCarlo Moschini, and Joseph Cooper. "Welfare Impacts of Alternative Biofuel and Energy Policies." *American Journal of Agricultural Economics* 93, no. 5 (2011): 1235–1256.

Dawe, David. "The Unimportance of Low World Grain Stocks for Recent World Price Increases." ESA Working Paper no. 09–91, 2009.

Dawe, David. *The Rice Crisis: Markets, Policies and Food Security*. London and Washington, DC: Earthscan, 2010.

Dawe, David, and Tom Slayton. "The World Rice Market Crisis of 2007–2008." In *The Rice Crisis: Markets, Polices and Food Security*, edited by David Dawe, 15–28. London and Washington, DC: Food and Agriculture Organization Rome and Earthscan Publishing, 2010.

Dawe, David, and Peter C. Timmer. "Why Stable Food Prices Are a Good Thing: Lessons from Stabilizing Rice Prices in Asia." *Global Food Security* 1, no. 2 (2012): 127–133.

de Gorter, Harry, and Dusan Drabik. "The Effect of Biofuel Policies on Food Grain Commodity Prices." *Biofuels* 3, no. 1 (2012a): 21–24.

de Gorter, Harry, and Dusan Drabik. "Biofuel Policies and Grain Crop Price Volatility." *Biofuels* 3, no. 2 (2012b): 111–113.

de Gorter, Harry, and Dusan Drabik. "Socioeconomic Issues and Biofuel Energy," Chapter 6. In *Our Energy Future: Socioeconomic Implications and Policy Options for Rural America*, edited by Don Albrecht. Routledge: Taylor & Francis, 2014.

de Gorter, Harry, and David R. Just. "The Welfare Economics of an Excise-Tax Exemption for Biofuels and the Interaction Effects with Farm Subsidies." Department of Applied Economics and Management Working Paper # 2007–13, Cornell University, September 2007a.

de Gorter, Harry, and David R. Just. "The Law of Unintended Consequences: How the U.S. Biofuel Tax Credit with a Mandate Subsidizes Oil Consumption and Has No Impact on Ethanol Consumption." Department of Applied Economics and Management Working Paper # 2007–20, Cornell University, October 2007b.

de Gorter, Harry, and David R. Just. "The Economics of U.S. Ethanol Import Tariffs with a Consumption Mandate and Tax Credit." Department of Applied Economics and Management Working Paper # 2007–21, Cornell University, October 2007c.

de Gorter, Harry, and David R. Just. "'Water' in the U.S. Ethanol Tax Credit and Mandate: Implications for Rectangular Deadweight Costs and the Corn-Oil Price Relationship." SSRN Working Papers, December 2007d.

de Gorter, Harry, and David R. Just. "The Economics of the U.S. Ethanol Import Tariff with a Blend Mandate and Tax Credit." *Journal of Agricultural & Food Industrial Organization* 6, no. 2 (2008a): 1–23.

de Gorter, Harry, and David R. Just. "'Water' in the U.S. Ethanol Tax Credit and Mandate: Implications for Rectangular Deadweight Costs and the Corn-Oil Price Relationship." *Review of Agricultural Economics* 30, no. 3 (2008b): 397–410.

de Gorter, Harry, and David R. Just. "The Welfare Economics of a Biofuel Tax Credit and the Interaction Effects with Price Contingent Farm Subsidies." *American Journal of Agricultural Economics* 91, no. 2 (2009a): 477–488.

de Gorter, Harry, and David R. Just. "The Economics of a Blend Mandate for Biofuels." *American Journal of Agricultural Economics* 91, no. 3 (2009b): 738–750.

de Gorter, Harry, and David R. Just. "Why Sustainability Standards for Biofuel Production Make Little Economic Sense." *Cato Institute Policy Analysis* 647 (2009c): 1–12.

de Gorter, Harry, and David R. Just. "The Social Costs and Benefits of Biofuels: The Intersection of Environmental, Energy and Agricultural Policy." *Applied Economic Perspectives and Policy* 32, no. 1 (2010a): 4–32.

de Gorter, Harry, and David R. Just. "Ethanol and Corn Prices: The Role of U.S. Tax Credits, Mandates and Import Tariffs," Chapter 9. In *The Economic Impact of Public Support to Agriculture: An International Perspective*, edited by Virgil Ball, Roberto Fanfani, and Luciano Gutierrez New York: Springer, 2010b.

de Gorter, Harry, Dusan Drabik, and David R. Just. "The Economics of a Blender's Tax Credit Versus a Tax Exemption: The Case of U.S. 'Splash & Dash' Biodiesel Exports to the European Union." *Applied Economic Perspectives and Policy* 33, no. 4 (2011): 510–527.

de Gorter, Harry, Dusan Drabik, and David R. Just. "Biofuel Policies and Food Grain Commodity Prices 2006–2012: All Boom and no Bust?" *AgBioForum* 16, no. 1 (2013a): 1–13.

de Gorter, Harry, Dusan Drabik, and David R. Just. "How Biofuels Policies Affect the Level of Grains and Oilseed Prices: Theory, Models and Evidence." *Global Food Security* 2, no. 2 (2013b): 82–88.

de Gorter, Harry, Dusan Drabik, and David R. Just. "Policy Implications of High RIN Prices and the 'Blend Wall'." *Biofuels* 4, no. 4 (2013c): 359–361.

de Gorter, Harry, Dusan Drabik, and David R. Just. "The Perverse Effects of Biofuel Public-Sector Policies." *Annual Review of Resource Economics* 5 (2013d): 463–483.

de Gorter, Harry, Dusan Drabik, David R. Just, and Erika M. Kliauga. "The Impact of OECD Biofuels Policies on Developing Countries." *Agricultural Economics* 44, no. 4 (2013e): 477–486.

de Gorter, Harry, Dusan Drabik, Erika M. Kliauga, and Govinda R. Timilsina. "An Economic Model of Brazil's Ethanol-Sugar Markets and Impacts of Fuel Policies." Policy Research Working Paper #6524, The World Bank, June 2013f.

de Gorter, Harry, Dusan Drabik, and Erika M. Kliauga. "The Role of World Biofuel Policies and Trade on the Level and Volatility of Grain/Oilseed Prices." In *Trade Policy and Food Security. Directions in Development Trade*, edited by Ian Gillson. Washington, DC: The World Bank, 2014.

de Nicola, Francesca, Pierangelo De Pace, and Manuel A. Hernandez. "Co-Movement of Major Commodity Price Returns Time-Series Assessment." Policy Research Working Paper 6845. The World Bank, Africa Region Finance and Private Sector Development Unit, April 2014.

De Schutter, Olivier. "Food Commodities Speculation and Food Price Crises: Regulation to Reduce the Risks of Price Volatility." United Nations Special Rapporteur on the Right to Food. Briefing note 02, September 2010. http://www.srfood.org/images/stories/pdf/otherdocuments/20102309_briefing_note_02_en_ok.pdf

Deaton, Angus, and Guy Laroque. "On the Behaviour of Commodity Prices." *Review of Economic Studies* 59, no. 1 (1992): 1–23.

Dillon, Brian M., and Christopher B. Barrett. "Global Oil Prices and Local Food Prices: Evidence from East Africa." A Working Paper, April 2014. http://dyson.cornell.edu/faculty_sites/cbb2/files/papers/DillonBarrett_140424.pdf

Dornbusch, Rudiger. "Expectations and Exchange Rate Dynamics." *Journal of Political Economy* 84, no. 6 (1976): 1161–1176.

Drabik, Dusan. "The Theory of Biofuel Policy and Food Grain Prices." Dyson School of Applied Economics and Management Working Paper # 2011–20, Cornell University, December 2011.

Drabik, Dusan, and Harry de Gorter. "Biofuels Policy, Drought and Corn Prices." *Biofuels* 3, no. 6 (2012): 649–651.

Drabik, Dusan, Harry de Gorter, and David R. Just. "The Implications of Alternative Biofuel Policies on Carbon Leakage." Dyson School of Applied Economics and Management Working Paper # 2010–22, Cornell University, November 2010.

Drabik, Dusan, Harry de Gorter, and Govinda R. Timilsina. "The Effect of Biodiesel Policies on World Biodiesel and Oilseed Prices." *Energy Economics* 44 (2014): 80–88.

Du, Xiaodong, and Dermot J. Hayes. "The Impact of Ethanol Production on U.S. and Regional Gasoline Markets: An Update to 2012." CARD Working Paper 12-WP 528, Iowa State University, May 2012. http://www.card.iastate.edu/publications/dbs/pdffiles/12wp528.pdf

The Economist. "Food and the Poor: The New Face of Hunger. Global Food Shortages Have Taken Everyone by Surprise. What Is to Be Done?" Last modified April 17, 2008a. http://www.economist.com/node/11049284

The Economist. "The Silent Tsunami: Food Prices Are Causing Misery and Strife around the World. Radical Solutions Are Needed." Last modified April 17, 2008b. http://www.economist.com/node/11050146

Energy Information Administration (EIA). "Annual Energy Outlook 2013 with Projections to 2040," April 2013. http://www.eia.gov/forecasts/aeo/pdf/0383%282013%29.pdf

Elam, Thomas E. "The RFS, Fuel and Food Prices, and the Need for Statutory Flexibility." FarmEcon LLC, July 16, 2012. http://www.nationalchickencouncil.org/wp-content/uploads/2012/07/RFS-issues-FARMECON-LLC-7-16-12-FINAL.pdf

Elobeid, Amani, and Simla Tokgoz. "Removing Distortions in the U.S. Ethanol Market: What Does It Imply for the United States and Brazil?" *American Journal of Agricultural Economics* 90, no. 4 (2008): 918–932.

Enders, Walter, and Matthew T. Holt. "Breaks, Bubbles, Booms, and Busts: The Evolution of Primary Commodity Price Fundamentals." MPRA Paper no. 3146, University of Alabama, June 12, 2011. http://mpra.ub.uni-muenchen.de/31461/

Enders, Walter, and Matthew T. Holt. "Sharp Breaks or Smooth Shifts? An Investigation of the Evolution of Commodity Prices." *American Journal of Agricultural Economics* 94, no. 3 (2012): 659–673.

Enders, Walter, and Matthew T. Holt. "The Evolving Relationships between Agricultural and Energy Commodity Prices: A Shifting-Mean Vector Autoregressive Analysis," Chapter 4. In *The Economics of Food Price Volatility*, edited by Jean-Paul Chavas, David Hummels, and Brian D. Wright. Chicago, IL: University of Chicago Press, October 2014.

Energy Independence and Security Act of 2007 (EISA). Department of Energy, United States Government. 2007.

Environmental Protection Agency (EPA). "EPA Proposes 2013 Renewable Fuel Standards." Office of Transportation and Air Quality. EPA-420-F-13-007, January 2013a. http://www.epa.gov/otaq/fuels/renewablefuels/documents/420f13007.pdf

Environmental Protection Agency (EPA). "Regulation of Fuels and Fuel Additives: 2013 Renewable Fuel Standards; Final Rule." *Federal Register* 78, no. 158 (August 2013b). http://www.gpo.gov/fdsys/pkg/FR-2013-08-15/pdf/2013-19557.pdf

Environmental Protection Agency (EPA). "Questions and Answers on Changes to the Renewable Fuel Standard Program (RFS2)." Accessed February 6, 2013c. http://www.epa.gov/otaq/fuels/renewablefuels/compliancehelp/rfs2-aq.htm#7

EU Commission. "Biofuels – Progress Reports." December 2011. http://ec.europa.eu/energy/renewables/biofuels/ms_reports_dir_2003_30_en.htm

EU Commission. "Renewable Energy Progress Report." March 2013. http://eur-lex.europa.eu/LexUriServ/LexUriServ.do?uri=COM:2013:0175:FIN:EN:PDF

FAPRI. "Biofuel Conversion Factors." Food and Agricultural Policy Research Institute, Missouri. Last accessed July 8, 2014. http://www.fapri.missouri.edu/outreach/publications/2006/biofuelconversions.pdf

Fischer, Carolyn, and Richard G. Newell. "Environmental and Technology Policies for Climate Mitigation." *Journal of Environmental Economics and Management* 55, no. 2 (2008): 142–162.

Fonseca Blanco, María, Alison Burrell, Hubertus Gay, Martin Henseler, Aikaterini Kavallari, Robert M'Barek, Ignácio Pérez Domínguez, and Axel Tonini. *Impacts of the EU Biofuel Target on Agricultural Markets and Land Use: A Comparative Modelling Assessment*. Luxembourg: Publications Office of the European Union, 2010.

Food and Agriculture Organization (FAO). "The State of Food and Agriculture. Biofuels: Prospects, Risks and Opportunities." A Report, 2008a. ftp://ftp.fao.org/docrep/fao/011/i0100e/i0100e.pdf

Food and Agriculture Organization. (FAO). "Soaring Food Prices: Facts, Perspectives, Impacts and Actions Required." Presentation at High-Level Conference on World Food Security: The Challenges of Climate Change and Bioenergy, Rome, Italy, June 3–5, 2008b.

Food and Agriculture Organization of the United Nations (FAO). "The State of Agricultural Commodity Markets: High Food Prices and the Food Crisis—Experiences and Lessons Learned." Rome: FAO, 2009.

Foster, Harry, Robert Baron, and Paul Bernstein. "Impact of the Blend Wall Constraint in Complying with the Renewable Fuel Standard." Charles Rivers Associates, Washington, DC, Report for the American Petroleum Industry, November 2, 2011.

Foucault, Michel. *The Order of Things*. Taylor & Francis, 2002.

Frankel, Jeffrey. "The Effect of Monetary Policy on Real Commodity Prices," Chapter 7. In *Asset Prices and Monetary Policy*, edited by John Campbell. Chicago, IL: University of Chicago Press, 2008.

Frankel, Jeffrey, and Andrew K. Rose. "Determinants of Agricultural and Mineral Commodity Prices." In *Inflation in an Era of Relative Price Shocks*, edited by Renée Fry, Callum Jones, and Christopher Kent. Sydney, NSW: Reserve Bank of Australia, 2009.

Galtier, Franck. "Managing Food Price Instability: Critical Assessment of the Dominant Doctrine." *Global Food Security* 2, no. 2 (2013): 72–81.

Gardebroek, Cornelis, and Manuel A. Hernandez. "Do Energy Prices Stimulate Food Price Volatility? Examining Volatility Transmission between US Oil, Ethanol and Corn Markets." *Energy Economics* 40 (2013): 119–129.

Gardebroek, Cornelis, Manuel A. Hernandez, and Miguel Robles. "Market Interdependence and Volatility Transmission among Major Crops." Presentation at the AAEA & CAES Joint Annual Meeting, Washington, DC, August 4–6, 2013.

Gardner, Bruce. "Fuel Ethanol Subsidies and Farm Price Support." *Journal of Agricultural and Food Industrial Organization* 5, no. 2 (2007): 1–22.

Gilbert, Christopher L. "How to Understand High Food Prices." *Journal of Agricultural Economics* 61, no. 2 (2010): 398–425.

Gilbert, Christopher L., and Wyn C. Morgan. "Food Price Volatility." *Philosophical Transactions of the Royal Society B* 365, no. 1554 (2010): 3023–3034.

Giordani, Paolo E., Nadia Rocha, and Michele Ruta. "Food Prices and the Multiplier Effect of Export Policy." WTO Working Paper ERSD-2012–08, April 2012. http://www.wto.org/english/res_e/reser_e/ersd201208_e.pdf

Guariso, Andrea, Mara Squicciarini, and Johan F.M. Swinnen. "Food Price Shocks and the Political Economy of Global Agricultural and Development Policy." LICOS Discussion Paper 348/2014, March 14, 2014. http://papers.ssrn.com/sol3/papers.cfm?abstract_id=2419377

Hamelinck, Carlo. "Biofuels and Food Security: Risks and Opportunities." Project no. BIENL13469. Ecofys. August 2013. http://www.ecofys.com/files/files/ecofys-2013-biofuels-and-food-security.pdf

Harri, Ardian, Lanier Nalley, and Darren Hudson. "The Relationship between Oil, Exchange Rates, and Commodity Prices." *Journal of Agricultural and Applied Economics* 41, no. 2 (2009): 501–510.

Headey, Derek, and Shenggen Fan. *Reflections on the Global Food Crisis: How It Happened? How It Hurt? And, How We Can Prevent the Next One?* Research Monograph 165. Washington, DC: International Food Policy Research Institute, 2010.

Hertel, Thomas W., and Jayson Beckman. "Commodity Price Volatility in the Biofuel Era: An Examination of the Linkage between Energy and Agricultural Markets," Chapter 6. In *The Intended and Unintended Effects of U.S. Agricultural and Biotechnology Policies*, edited by Graff Zivin, Joshua S., and Jeffrey M. Perloff. National Bureau of Economic Research, University of Chicago Press, 2012.

HLPE. "Price Volatility and Food Security." A report by the High Level Panel of Experts on Food Security and Nutrition of the Committee on World Food Security, Rome, July, 2011.

HLPE. "Biofuels and Food Security." A report by the High Level Panel of Experts on Food Security and Nutrition of the Committee on World Food Security, Rome, June 2013.

Hochman, Gal, Deepak Rajagopal, Govinda Timilsina, and David Zilberman. "The Role of Inventory Adjustments in Quantifying Factors Causing Food Price Inflation." Policy Research Working Paper 5744, Development Research Group, Environment and Energy Team. Washington, DC: World Bank, August 2011.

Holland, Stephen P., Jonathan E. Hughes, Christopher R. Knittel, and Nathan C. Parker. "Some Inconvenient Truths About Climate Change Policy: The Distributional Impacts of Transportation Policies." *Review of Economics and Statistics* (2014, forthcoming).

Huchet-Bourdon, Mariline. "Agricultural Commodity Price Volatility: An Overview." OECD Food, Agriculture and Fisheries Working Papers, no. 52, OECD Publishing, 2011.

Interagency Report to the G20. FAO, IFAD, IMF, OECD, UNCTAD, WFP, the World Bank, the WTO, IFPRI, and the UN HLTF. "Price Volatility in Food and Agricultural Markets: Policy Responses." Policy Report for the G-20, June 2, 2011.

Intergovernmental Panel on Climate Change (IPCC). "2006 IPCC Guidelines for National Greenhouse Gas Inventories." April 2007. http://www.ipcc-nggip.iges.or.jp/public/2006gl/

Irwin, Scott H., and Darrel Good. "Ethanol – Does the RFS Matter?" Farmdoc Daily, University of Illinois. University of Illinois, August 2, 2012. http://www.farmdocdaily.illinois.edu/2012/08/ethanoldoes_the_rfs_matter.html

Irwin, Scott H., and Darrel Good. "Why Are Ethanol Prices and Production Profits So High?" Farmdoc Daily, University of Illinois. University of Illinois, June 11, 2014. http://farmdocdaily.illinois.edu/2014/06/why-are-ethanol-prices-and-production-profits-so-high.html

Irwin, Scott H., and Dwight R. Sanders. "Index Funds, Financialization, and Commodity Futures Markets.' *Applied Economic Perspectives and Policy* 33, no. 1 (2011): 1–31.

Ivanic, Maros, and Will Martin. "Implications of Domestic Price Insulation for Global Food Price Behavior." *Journal of International Money and Finance* 42 (2014a): 272–288.

Ivanic, Maros, and Will Martin. "World Food Price Rises and the Poor 2006–12: A Slow Food Price Crisis?" World Bank Working Paper, June 9, 2014b.

Janzen, Joseph P., Colin Carter, Aaron Smith, and Michael K. Adjemian. "Deconstructing Wheat Price Spikes: A Model of Supply and Demand, Financial Speculation, and Commodity Price Comovement." ERR-165, U.S. Department of Agriculture, Economic Research Service, April 2014.

Jones, Darryl, and Andrzej Kwiecinski. "Policy Responses in Emerging Economies to International Agricultural Commodity Price Surges." OECD Food, Agriculture and Fisheries Papers no. 34, 2010. http://dx.doi.org/10.1787/5km6c61fv40w-en

Kauffman, Nathan. "Markets, Not Mandates, Shape Ethanol Production." *The Main Street Economist: Agricultural and Rural Analysis* no. 5 (2012): 1–7. http://www.kc.frb.org/publicat/mse/MSE_0512.pdf

Kliauga, Erika M., Harry de Gorter, and David R. Just. "Measuring the Subsidy Component of Biofuel Tax Credits and Exemptions." In *The Economics of Alternative Energy Sources and Globalization*, edited by Andrew Schmitz, Norbert L. Wilson, Charles B. Moss, and David Zilberman, 49–64. United Arab Emirates: Bentham Science Publishers, 2011.

Knittel, Christopher R., and Aaron Smith. "Ethanol Production and Gasoline Prices: A Spurious Correlation." *The Energy Journal* 36, no. 1(2015).

Konandreas, Panos. "Trade Policy Responses to Food Price Volatility in Poor Net Food-Importing Countries." ICTSD Programme on Agricultural Trade and Sustainable Development Issue Paper no. 42, June 2012.

Koplow, Doug. "State and Federal Subsidies to Biofuels: Magnitude and Options for Redirection." *International Journal of Biotechnology* 11, no. 1–2 (2009): 92–126.

Kristoufek, Ladislav, Karel Janda, and David Zilberman. "Correlations between Biofuels and Related Commodities before and during the Food Crisis: A Taxonomy Perspective." *Energy Economics* 34, no. 5 (2012): 1380–1391.

Lapan, Harvey, and GianCarlo Moschini. "Second-Best Biofuel Policies and the Welfare Effects of Quantity Mandates and Subsidies." *Journal of Environmental Economics and Management* 63, no. 2 (2012): 224–241.

Lewis, Michael. *Flash Boys*. New York: W.W. Norton & Company, 2014.

McPhail, Lihong Lu, and Bruce A. Babcock. "Ethanol, Mandates, and Drought: Insights from a Stochastic Equilibrium Model of the U.S. Corn Market." A working paper, March 2008.

Mallory, Mindy L., Scott H. Irwin, and Dermot J. Hayes. "How Market Efficiency and the Theory of Storage Link Corn and Ethanol Markets." *Energy Economics* 34, no. 6 (2012): 2157–2166.

McCalla, Alex F. "World Food Prices: Causes and Consequences." *Canadian Journal of Agricultural Economics* 57, no. 1 (2009): 23–34.

McPhail, Lihong, Paul Westcott, and Heather Lutman. "The Renewable Identification Number System and U.S. Biofuel Mandates." USDA Economic Research Service, BIO-03 November, 2011.

McPhail, Lihong Lu, and Bruce A. Babcock. "Impact of US Biofuel Policy on US Corn and Gasoline Price Variability." *Energy* 37, no. 1 (2012): 505–513.

Meyer, Seth, and Rob Johansson. "Proposed Rules for 2014 RFS – A Volume of Uncertainty." Farmdoc Daily, University of Illinois. University of Illinois, November 20, 2013. http://farmdocdaily.illinois.edu/2013/11/proposed-rules-2014-rfs.html

Meyer, Seth, Josef Schmidhuber, and Jesús Barreiro-Hurlé. "Intra-Industry Trade in Biofuels: How Environmental Legislation Fuels Resource Use and GHG Emissions." ESA Working Paper no. 12–08, Agricultural Development Economics Division, Food and Agriculture Organization, 2012. http://www.fao.org/docrep/017/ap869e/ap869e.pdf

Mitchell, Donald. "A Note on Rising Food Prices." Policy Research Working Paper no. 4682, Development Prospects Group. Washington, DC: World Bank, July 2008. http://papers.ssrn.com/sol3/papers.cfm?abstract_id=1233058

Moschini, GianCarlo, Jingbo Cui, and Harvey Lapan. "Economics of Biofuels: An Overview of Policies, Impacts and Prospects." *Bio-Based and Applied Economics* 1, no. 3 (2012): 269–296.

Msangi, Siwa, and Martin Evans. "Biofuels and Developing Economies: Is the Time Right?" *Agricultural Economics* 44, no. 4–5 (2013): 501–510.

Myers, Robert J., Stanley R. Johnson, Michael Helmar, and Harry Baumes. "Long-Run and Short-Run Co-Movements in Energy Prices and the Prices of Agricultural Feedstocks for Biofuel." *American Journal of Agricultural Economics* 96, no. 4 (2014): 991–1008.

National Academy of Sciences (NAS). *Renewable Fuel Standard: Potential Economic and Environmental Effects of U.S. Biofuel Policy*. Washington, DC: National Academies Press, 2011.

Nuffield Council on Bioethics. "Biofuels: Ethical Issues." April, 2011. http://nuffieldbioethics.org/wp-content/uploads/2014/07/Biofuels_ethical_issues_FULL-REPORT_0.pdf

OECD. "Biofuel Support Policies: An Economic Assessment." Paris, 2008. http://www.oecd.org/fr/tad/echanges-agricoles/biofuelsupportpoliciesaneconomicassessment.htm

OECD. "Agricultural Policy Monitoring and Evaluation 2013." Paris, September 2013. http://www.oecd-ilibrary.org/agriculture-and-food/agricultural-policy-monitoring-and-evaluation-2013_agr_pol-2013-en;jsessionid=9usa6s89t6lu3.x-oecd-live-01

OECD/FAO. "Biofuels," Chapter 3. In *OECD-FAO Agricultural Outlook 2014–2023*. OECD Publishing, 2014.

Official Journal of the European Union. "EU Council Regulation (EC) No 598/2009 of 7 July 2009 imposing a definitive countervailing duty and collecting definitively the provisional duty imposed on imports of biodiesel originating in the United States of America." Julay 2009. http://eur-lex.europa.eu/LexUriServ/LexUriServ.do?uri=OJ:L:2009:179:0001:0025:EN:PDF

Oxfam. "Another Inconvenient Truth: How Biofuel Policies Are Deepening Poverty and Accelerating Climate Change." Oxfam Briefing Paper 114, June

2008. http://www.oxfam.org/sites/www.oxfam.org/files/bp114-inconvenient-truth-biofuels-0806.pdf

Paarlberg, Robert. *Food Politics: What Everyone Needs to Know*. Oxford University Press, 2010.

Peterson, E. Wesley F. *A Billion Dollars a Day: The Economics and Politics of Agricultural Subsidies*. Malden, MA: Wiley-Blackwell, 2009.

Pinstrup-Andersen, Per. "The Political Economy of Food Price Policy." Lecture in the Cornell Institute for Food and Agricultural Development seminar series, March 6, 2013a. http://vimeo.com/63107499

Pinstrup-Andersen, Per. "Don't Believe the Hype (and Data) Surrounding Food Price Crises Cites Food Price Volatility as Greater Danger than High Food Prices." *IFPRI blogs*. Last modified May 10, 2013b. http://www.ifpri.org/blog/cornell-s-pinstrup-andersen-don-t-believe-hype-and-data-surrounding-food-price-crises

Piot-Lepetit, Isabelle, and Robert M'Barek. "Methods to Analyze Agricultural Commodity Price," Chapter 1. In *Methods to Analyze Agricultural Commodity Price*, edited by Isabelle Piot-Lepetit, and Robert M'Barek. New York: Springer, 2011.

Prakash, Adam (ed). *Safeguarding Food Security in Volatile Global Markets*. Rome: FAO, 2011.

Qiu, Cheng, Gregory Colson, Cesar Escalante, and Michael Wetzstein. "Considering Macroeconomic Indicators in the Food before Fuel Nexus." *Energy Economics* 34, no. 6 (2012): 2021–2028.

Rajagopal, Deepak, Steven E. Sexton, Gal Hochman, David Roland-Holst, and David Zilberman. "Model Estimates Food Versus Biofuel Tradeoff." *California Agriculture* 63, no. 4 (2009): 199–201.

Rajcaniova, Miroslava, Dusan Drabik, and Pavel Ciaian. "How Policies Affect International Biofuel Price Linkages." *Energy Policy* 59 (2013): 857–865.

Rausser, Gordon C., and Harry de Gorter. "US Policy Contributions to Food Grain Commodity Prices." Presentation at UNU-Wider Workshop on The Political Economy of Food Price Policy, Cornell University, Ithaca NY, July 9–12, 2012. http://www.wider.unu.edu/publications/working-papers/2013/en_GB/wp2013–033/

Roberts, Michael J., and Wolfram Schlenker. "World Supply and Demand of Food Commodity Calories." *American Journal of Agricultural Economics* 91, no. 5 (2009): 1235–1242.

Roberts, Michael J., and Wolfram Schlenker. "Identifying Supply and Demand Elasticities of Agricultural Commodities: Implications for the US Ethanol Mandate." *American Economic Review* 103, no. 6 (2013): 2265–2295.

Roberts, Michael J., and Nam A. Tran. "Commodity Price Adjustment in a Competitive Storage Model with an Application to the US Biofuel Policies." Paper presented at the Agricultural & Applied Economics Associations 2012 AAEA Annual Meeting, Seattle, Washington, DC, August 12–14, 2012.

Rosegrant, Mark W., Tingju Zhu, Siwa Msangi, and Timothy Sulser. "Global Scenarios for Biofuels. Impacts and Implications." *Review of Agricultural Economics* 30, no. 3 (2008): 495–505.

Robles, Miguel, Maximo Torero, and Joachim von Braun. "When Speculation Matters." International Food Policy Research Institute, 2009.

Rutten, Martine, Lindsay Shutes, and Gerdien Meijerink. "Sit Down at the Ball Game: How Trade Barriers Make the World Less Food Secure." *Food Policy* 38, no. 1 (2013): 1–10.

Runge, Ford C., and Benjamin Senauer. "How Biofuels Could Starve the Poor." *Foreign Affairs* 86, no. 3 (May/June, 2007): 41–53.

Salvo, Alberto, and Cristian Huse. "Build It, But Will They Come? Evidence from Consumer Choice between Gasoline and Sugarcane Ethanol." *Journal of Environmental Economics and Management* 66, no. 2 (2013): 251–279.

Sandoval, Israel Solorio, and Lucia Popartan. "The Implementation of the EU Biofuels Policy in Spain and the UK: A Case of Contested Europeanization." *Biofuels* 5, no. 2 (2014): 129–140.

Searchinger, Timothy. "Understanding the Biofuel Trade-offs between Indirect Land Use Change (ILUC), Hunger and Poverty." Independent analysis produced for Friends of the Earth Europe, July, 2013.

Sen, Amartya. "Food, Economics, and Entitlements." In *International Agricultural Development*, edited by Carl K. Eicher and John M. Staatz, 240–258. Baltimore, MD: Johns Hopkins University Press, 1998.

Serra, Teresa. "Volatility Spillovers between Food and Energy Markets: A Semiparametric Approach." *Energy Economics* 33 (2011): 1155–1164.

Serra, Teresa, and David Zilberman. "Biofuel-Related Price Transmission Literature: A Review." *Energy Economics* 37 (2013): 141–151.

Serra, Teresa, David Zilberman, and Jose M. Gil. "Price Volatility in Ethanol Markets." *European Review of Agricultural Economics* 38, no. 2 (2011a): 259–280.

Serra, Teresa, David Zilberman, Jose M. Gil, and Barry K. Goodwin. "Nonlinearities in the U.S. Corn-Ethanol-Oil-Gasoline Price System." *Agricultural Economics* 42, no. 1 (2011b): 35–45.

Sexton, Steven E., and David Zilberman. "Land for Food and Fuel Production: The Role of Agricultural Biotechnology." In *The Intended and Unintended Effects of U.S. Agricultural and Biotechnology Policies*, edited by Joshua S. Graff Zivin and Jeffrey M. Perloff, 269–288. National Bureau of Economic Research, University of Chicago Press, 2012.

Sharma, Ramesh. "Food Export Restrictions: Review of the 2007–2010 Experience and Considerations for Disciplining Restrictive Measures." FAO Commodity and Trade Policy Research Working Paper no. 32. Rome: FAO, 2011. http://www.ictsd.org/downloads/2011/05/sharma-export-restrictions.pdf

Shepherd, Justin S. "The Complex Structure of the U.S. Biofuel Mandate and Implications for World Biofuel and Grain/Oilseed Prices." Master's Thesis, Charles H. Dyson School of Applied Economics and Management, Cornell University, 2014.

Shiller, Robert J. "Commodities Followed the Euphoria Cycle that We Had along with Housing." Yale University, New York Times, October 13, 2008. http://www.nytimes.com/2008/10/14/business/economy/14commodities.html?_r=2&oref=slogin&

Shinoj, P., S.S. Raju, and P.K. Joshi. "India's Biofuel Production Programme: Need for Prioritizing the Alternative Options." *Indian Journal of Agricultural Sciences* 81, no. 5 (2011): 391–397.

Sinn, Hans-Werner. "Public Policies against Global Warming: A Supply-Side Approach" *International Tax and Public Finance* 15, no. 4 (2008): 360–394.

Slayton, Tom. "Arson Forensics: What Set the World Rice Market on Fire in 2008?" Presentation at the FAO Workshop on Rice Policies in Asia, Chiang Mai, Thailand, February 9–12, 2009.

Slayton, Tom. "The 'Diplomatic Crop,' or How the US Provided Critical Leadership in Ending the Rice Crisis," Chapter 16. In *The Rice Crisis: Markets, Polices and Food Security*, edited by David Dawe. London and Washington, DC: Food and Agriculture Organization Rome and Earthscan Publishing, 2010.

Smith, Aaron. "Giving an Inch and Keeping a Mile: Why the Corn Lobby Let the Ethanol Tax Credit Expire." *Agricultural and Resource Economics Update* 15, no. 5 (2012): 1–4.

Steenblik, Ronald. "Biofuels – At what cost?: Government Support for Ethanol and Biodiesel in Selected OECD Countries." Global Subsidies Initiative, September, 2007.

Stoeckel, Andy. "High Food Prices: Causes, Implications and Solutions." Publication no. 08/100, Rural Industries Research and Development Corporation, Australian Government, June 2008.

Swinnen, Johan, and Pasquamaria Squicciarini. "Mixed Messages on Prices and Food Security." *Science* 335, no. 6067 (2012): 405–406.

Switzer, Stephanie, and Joseph A. McMahon. "EU Biofuels Policy – Raising the Question of WTO Compatibility." *International and Comparative Law Quarterly* 60, no. 3 (2011): 713–736.

Taleb, Nassim Nicholas. *Antifragile: Things That Gain from Disorder*. New York: Random House, 2012.

Takarick, Stephen. "Dispelling Some Misconceptions about Agricultural Trade Liberalization." *Journal of Economic Perspectives* 22, no. 1 (2008): 199–216.

Tangermann, Stefan. "Policy Solutions to Agricultural Market Volatility: A Synthesis." ICTSD Programme on Agricultural Trade and Sustainable Development, Issue Paper no. 33, June 2011.

Thompson, Wyatt, Seth Meyer, and Pat Westhoff. "The New Markets for Renewable Identification Numbers." *Applied Economic Perspectives and Policy* 32, no. 4 (2010): 588–603.

Timilsina, Govinda R., and Ashish Shrestha. "Biofuels: Markets, Targets and Impacts." World Bank Policy Research Working Paper 5364, July, 2010.

Timmer, Peter C. "Causes of High Food Price." ADB Economics Working Paper no. 128, Asian Development Bank, Manila, October 2008.

Timmer, Peter C. "Did Speculation Affect World Rice Prices?" ESA Working Paper no. 09–07, April 2009.

Timmer, Peter C. "Reflections on Food Crises Past." *Food Policy* 35, no. 1 (2010): 1–11.

Timmer, Peter C., and David Dawe. "Food Crises Past, Present (and Future?): Will We Ever Learn?" In *The Rice Crisis: Markets, Polices and Food Security*, edited by

David Dawe, 3–14. London and Washington, DC: Food and Agriculture Organization and Earthscan Publishing, 2010.

Tomei, Julia, and Paul Upham. "Argentinean Soy-Based Biodiesel: An Introduction to Production and Impacts." *Energy Policy* 37, no. 10 (2009): 3890–3898.

Tomek, William G., and Harry M. Kaiser. *Agricultural Product Prices*, 5th edn. Ithaca, NY: Cornell University Press, 2014.

Trostle, Ronald. "Global Agricultural Supply and Demand: Factors Contributing to the Recent Increase in Food Commodity Prices." WRS-0801, Economic Research Service, U.S. Department of Agriculture, May 2008.

Trostle, Ronald, Daniel Marti, Stacey Rosen, and Paul Westcott. "Why Have Food Commodity Prices Risen Again?" WRS-1103, USDA, Economic Research Service, June 2011.

Tyner, Wallace E. "The US Ethanol and Biofuels Boom: Its Origins, Current Status, and Future Prospects." *BioScience* 58, no. 7 (2008): 646–653.

Tyner, Wallace E. "Biofuel Economics and Policy in the US." In *Biofuels Economics and Policy. Agricultural and Environmental Sustainability*, edited by Adele Finco, 41–57. Milan: FrancoAngeli, 2012.

Tyner, Wallace E. "The Future of Biofuels?" Presentation at the Sixth Annual Berkeley Bioeconomy Conference. College of Natural Resources and EBI, Berkeley, CA, March 27–28, 2013. http://www.berkeleybioeconomy.com/wp-content/uploads//2013/04/The-Future-of-Biofuels.pdf

Tyner, Wallace E., and Daniela Viteri. "Implications of Blending Limits on the US Ethanol and Biofuels Markets." *Biofuels* 1, no. 2 (2010): 251–253.

Tyner, Wallace E., Farzad Taheripour, and Christopher Hurt. "Potential Impacts of a Partial Waiver of the Ethanol Blending Rules." Farm Foundation, August 2012.

UFOP. "Environmental Committee rules in favour of Decline in European Biofuel Production." July 2013. http://www.ufop.de/english/news/environmental-committee-rules-in-favour-of-decline-in-european-biofuel-production/

United States Department of Agriculture (USDA). "EU Biofuels Annual 2013." GAIN Report no. NL3034, August 13, 2013.

United States Department of Agriculture (USDA). "World Agricultural Supply and Demand Estimates Report (WASDE)." Various issues. http://www.usda.gov/oce/commodity/wasde/

Wetzstein, Michael, and H. Wetzstein. "Four Myths Surrounding U.S. Biofuels." *Energy Policy* 39, no. 7 (2011): 4308–4312.

Williams, Jeffrey C., Brian D. Wright. *Storage and Commodity Markets*. Cambridge, UK: Cambridge University Press, 1991.

Wright, Brian D. "Speculators, Storage, and the Price of Rice." *Agricultural and Resource Economics Update* 12, no. 2 (2008): 7–10.

Wright, Brian D. "The Economics of Grain Price Volatility." *Applied Economic Perspectives and Policy* 33, no. 1 (2011): 32–58.

Wright, Brian D. "Global Biofuels: Key to the Puzzle of Grain Market Behavior." *Journal of Economic Perspectives* 28, no. 1 (2014): 73–98.

Yano, Yuki, David Blandford, and Yves R. Surry. "The Impact of Feedstock Supply and Petroleum Price Variability on Domestic Biofuel and Feedstock

Markets – The Case of the United States." Swedish University of Agricultural Sciences (SLU) Working Paper 2010/3, 2010.

Yano, Yuki, David Blandford, and Yves Surry. "Alternative Domestic and Trade Policies for Biofuels and Market Variability in the United States." In *The Economics of Alternative Energy Sources and Globalization*, edited by Andrew Schmitz, Norbert L. Wilson, Charles B. Moss, and David Zilberman, 49–64. United Arab Emirates: Bentham Science Publishers, 2011.

Yoon, Jesse Jin. "What's the Difference between Biodiesel and Renewable (Green) Diesel." Report prepared for Advanced Biofuels USA, 2011. http://advancedbiofuelsusa.info/wp-content/uploads/2011/03/11–0307-Biodiesel-vs-Renewable_Final-_3_-JJY-formatting-FINAL.pdf

Zhang, Zibin, Luanne Lohr, Cesar Escalante, and Michael Wetzstein. "Ethanol, Corn, and Soybean Price Relations in a Volatile Vehicle-Fuels Market." *Energies* 2, no. 2 (2009): 320–339.

Zhang, Zibin, Luanne Lohr, Cesar Escalante, and Michael Wetzstein. "Food Versus Fuel: What Do Prices Tell Us?" *Energy Policy* 38, no. 1 (2010): 445–451.

Zhang, Wei, Elaine A. Yu, Scott Rozelle, Jun Yang, and Siwa Msangi. "The Impact of Biofuel Growth on Agriculture: Why is the Range of Estimates So Wide?" *Food Policy* 38 (2013): 227–239.

Zilberman, David, Gal Hochman, Deepak Rajagopal, Steven E. Sexton, and Govinda R. Timilsina. "The Impact of Biofuels on Commodity Food Prices: Assessment of Findings." *American Journal of Agricultural Economics* 95, no. 2 (2013): 275–281.

DATA SOURCES

AMRC (Agricultural Marketing Resource Center). "Ethanol, Corn and DDGS Prices at Production Facility by State." Ann Johanns, Iowa State University—updated monthly, 2014. http://www.agmrc.org/renewable_energy/ethanol/ethanol-prices-trends-and-markets/

Bloomberg, L.P. "Commodity Price Graph for 1/1/06 to 1/30/2014." *Bloomberg Database*. Ithaca, NY: Cornell University. Accessed June 10, 2014.

British Petroleum. "Statistical Review of World Energy." June 2013. http://www.bp.com/statisticalreview

CARD. "Historical Ethanol Operating Margins." Spreadsheet of Iowa State University, 2014a. http://www.card.iastate.edu/research/bio/tools/hist_eth_gm.aspx

CARD. "Historical Biodiesel Operating Margins." Spreadsheet of Iowa State University, 2014b. http://www.card.iastate.edu/research/bio/tools/hist_eth_gm.aspx

EIA (Energy Information Administration). "U.S. Gulf Coast Conventional Gasoline Regular Spot Price FOB." 2014a. http://www.eia.gov/dnav/pet/hist/LeafHandler.ashx?n=pet&s=eer_epmru_pf4_rgc_dpg&f=m

EIA. "Renewable and Alternative Fuels Database." Accessed June 6, 2014b. http://www.eia.gov/renewable/reports.cfm?t=9999&f=d

EIA. "Spot Prices." 2014c. http://www.eia.gov/dnav/pet/pet_pri_spt_s1_m.htm

EPA (Environmental Protection Agency). "40 CFR Part 80: Regulation of Fuels and Fuel Additives: 2013 Renewable Fuel Standards; Final Rule. Federal Register." 2013.

EPA. "2013 RFS2 Data." Accessed June 6, 2014. http://www.epa.gov/otaq/fuels/rfsdata/2013emts.htm

GaspriceWatch.com. "Federal and State Fuel Taxes Gasoline and Diesel in Cents Per Gallon." 2014. http://www.gaspricewatch.com/web_gas_taxes.php

Index Mundi. "Soybeans, U.S. No. 2 Yellow, FOB Gulf of Mexico, U.S. Price, US$ Per Metric Ton" and "China Soybean Oilseed Imports by Year (1,000 Metric Tonnes)." 2014. http://www.indexmundi.com/commodities/?commodity=soybeans&months=60 and http://www.indexmundi.com/agriculture/?country=cn&commodity=soybean-oilseed&graph=imports

OPIS (Oil Price Information Service). Purchased Data.

S&P/Case-Shiller 20-City Composite Home Price Index. http://us.spindices.com/indices/real-estate/sp-case-shiller-20-city-composite-home-price-index

UFOP (Union zur Förderung von Oel- und Proteinpflanzen). "UFOP-Market Information Oilseeds and Biofuels." 2014. http://www.ufop.de/medien/downloads/english/market-information/

US Government Printing Office. "Electronic Code of Federal Regulations: Subpart M-Renewable Fuel Standard." Accessed June 16, 2014. http://www.ecfr.gov/cgi-bin/text-idx?c=ecfr&SID=27e92597791f3c5cb9d729f136e9782d&rgn=div6&view=text&node=40:17.0.1.1.9.13&idno=40#40:17.0.1.1.9.13.93.4

USDA Agricultural Exchange Rate Data Set. "Real Monthly Commodity Trade Weighted Exchange Rates. U.S. Markets Agricultural Trade." http://www.ers.usda.gov/data-products/agricultural-exchange-rate-data-set.aspx#.VBiQBxAXOMs

USDA Feed Grain Yearbook Table 12. "No. 2 Yellow, Central IL." http://www.ers.usda.gov/data-products/feed-grains-database/feed-grains-yearbook-tables.aspx#.U0_LKxAvB3s

USDA Rice Yearbook Table 20. "Thailand Milled Rice Prices, f.o.b. Bangkok A1 Super." http://www.ers.usda.gov/data-products/rice-yearbook-2014.aspx#.U0_MQhAvB3s

USDA Wheat Yearbook Table 20. "No. 1 Hard Red Winter, Gulf." http://www.ers.usda.gov/data-products/wheat-data.aspx#.VBhq8hAXOMt

World Bank. "Prices (Pink Sheet) Monthly World Prices of Commodities and Indices." 2014. http://econ.worldbank.org/WBSITE/EXTERNAL/EXTDEC/EXTDECPROSPECTS/0,,contentMDK:21574907~menuPK:7859231~pagePK:64165401~piPK:64165026~theSitePK:476883,00.html

INDEX

Printed and bound in the United States of America